NMR

Basic Principles and Progress
Grundlagen und Fortschritte

Volume 10

Van der Waals Forces and Shielding Effects

Editors: P. Diehl   E. Fluck   R. Kosfeld

Springer-Verlag

Berlin Heidelberg New York 1975

Professor Dr. P. DIEHL

Physikalisches Institut der Universität Basel

Professor Dr. E. FLUCK

Institut für Anorganische Chemie

der Universität Stuttgart

Professor Dr. R. KOSFELD

Institut für Physikalische Chemie

der Rhein.-Westf. Technischen Hochschule Aachen

With 13 Figures

ISBN-13: 978-3-642-66178-5   e-ISBN-13: 978-3-642-66176-1
DOI: 10.1007/978-3-642-66176-1

Library of Congress Cataloging in Publication Data: Rummens, Frans H A 1933 – Van der Waals forces in NMR intermolecular shielding effects. (NMR basic principles and progress; v. 10) Bibliography: p. Includes index. 1. Van der Waals forces. 2. Nuclear magnetic resonance spectroscopy. 3. Molecules. I. Title. II. Series. QC490.N2 vol. 10 [QD461]538'.3s [541'.226] 75-15821

Van der Waals Forces in NMR Intermolecular Shielding Effects

Frans H. A. Rummens

Department of Chemistry
University of Regina
Regina, Saskatchewan S4S OA2, Canada

Contents

Contents

2

Introduction and Foreword

The usefulness of solvent effect studies on NMR chemical shifts need not be elaborated here; many applications of solvent effects continue to be published in great profusion. Quite a few intermolecular phenomenae may contribute to solvent shifts, but there is always the ubiquitous Van der Waals effect σ_w. Contrary to such other effects as neighbour anisotropy σ_a, reaction field contribution σ_E or complexation effects σ_c, no major direct use has yet been found for the Van der Waals effect. So far the role of the Van der Waals effect has been that of a nasty, disturbing phenomenon, something to be eliminated at all costs. But it is precisely in this latter respect where almost all solvent effect studies fall short. Not only is σ_w usually large (larger than σ_a and σ_E even in ^{1}H NMR and probably the dominating term with heavier nuclei), but it is strongly variable from one solute to another and even from one nuclear site to another in the same solute molecule. No referencing technique, however cleverly devised, will be capable of eliminating the σ_w contribution from the other, presumedly more interesting contributions.

It appeared quite recently that mathematical trickery by the name of "factor analysis" could achieve the sought-for separation of contribuants. Our analysis of this technique, as detailed in Chapter 12, boils down to the following lemma; "Factor analysis can, in principle, separate physically different contribuants to the total solvent effect but in order to achieve this, it is essential to know intimately the physical laws that govern these various contribuants; under such conditions, however, one possesses already the knowledge one wants to obtain, and the subsequent factor analysis reveals nothing further of interest".

There appears to be only one possibility left and that is to develop models to *calculate* σ_w in any given circumstance. This approach essentially fills the pages of this review. Many models, each with many more refinements, have been proposed. Such models can be tested out on solvent shifts where σ_w is the *only* term (i.e. non-polar solutes in non-polar isotropic solvents). Once quantitative agreement between calculation and experiment can be obtained for a wide range of physical and chemical conditions, such a model could then safely be used to calculate σ_w in systems where more than one solvent effect term is present. In the 15 years or so that have elapsed since the discovery of the σ_w effect, PROGRESS toward this end has been made; yet, the picture is far from complete. It has turned out that the σ_w effect is very complex in nature and that many physical and molecular parameters must be considered before a quantitative understanding may be expected. In spite of this and partly because of this complex and difficult nature the study of σ_w should be continued, not only because of the nuisance character of σ_w, particularly with heavier nuclei, but also because a rich positive yield may be expected. Van der Waals shielding effects are site-specific sensors of intermolecular forces and should therefore be useful as a direct tool in structural or dynamic studies. Furthermore, σ_w offers possibilities for the study of intermolecular potentials and "structures" of liquids and gases that are unique in comparison to other physical techniques. It is hoped that this review will serve as a stimulant towards exploring these possibilities.

The term "Van der Waals shift" is not always used unambiguously. In this review σ_w will refer to the effect on the shielding due to dispersion forces or London forces including higher order dispersion terms if appropriate and also to the contributions to the shielding due to repulsive or overlap forces. In short, this review will deal only with non-polar solutes in non-polar magnetically isotropic solvents. With "non-polar" the absence of a permanent electric dipole moment is meant; this eliminates a solvent like $C(OMe)_4$ but a solvent like dioxane (whose first non-zero electric moment is a quadrupole) is included. The solvents CS_2 and $C(NO_2)_4$, although non-polar, are rather rigorously excluded, because they may have a neighbour anisotropy effect. For solutes the situation is not so clear; protons residing in C–H bonds of non-polar solutes might still have a reaction field contribution because the bond in which the proton resides is slightly polar. It is assumed, however, that this electric field effect is negligible for such protons. On the other hand, for the ^{19}F signals of *para*-$C_6H_4F_2$, although included in this review, such approximation may be quite wrong; the molecule as a whole is non-polar but the nucleus in question resides in the strongly polar C–F bonds. The same may be true for non-polar solutes like CF_4, SiF_4 and SF_6.

A further restriction was imposed quite deliberately by eliminating, with few exceptions, the discussion of data or experiments performed without reference to the gas shift of the solute. In such studies, whether they utilise internal referencing or external referencing to some other signal, much of the BASIC PRINCIPLES of the σ_w effect remain hidden. Such difference data $\Delta\sigma_w$ may be quite useful in practice, but the basic data and relations as given in this review should serve to make interpretations of $\Delta\sigma_w$ data meaningful; the inverse relation is less fruitful.

On a few occasions we have strayed from the above noted restrictions. For example, it seemed difficult to give a meaningful discussion of the temperature dependence of σ_w without referring to the problems concerning the intrinsic temperature dependence of shielding parameters. Similarly, a discussion of factor analysis and its importance for the extraction of σ_w is virtually impossible without delving into some of the finer points regarding other medium shift contribuants. Furthermore we wanted to provide a practical guide for those who are interested in evaluating σ_w in systems other than those discussed here. Hence the inclusion of such sections as on experimental techniques, on the bulk susceptibility term σ_b, on estimating required molecular parameters and on calculating σ_w.

Finally, this review is unlike most other contributions, because of its unhampered style and lack of objectivity. The reader may find the treatment of certain theories or certain results subjective and biased; he may find the statements speculative, contrived, opinionated or even downright erroneous.

The author wanted to write a *critical* review, but in his opinion all criticism is subjective. This review has been written by someone who has spent much effort on problems concerning σ_w, its origins and its applications. In a strictly objective review much of that personal experience would be lost; the review might just as well have been written by a competent librarian. This attitude and concomitant style admittedly carry some grave risks such as the wrath of those authors who may feel offended and the low regard of those who may find fault in this work. One thing is certain; a style as employed here is bound to raise counter criticism. So much the better. It is the author's believe that progress without strife is a contradiction. It is

also his growing conviction that science and scientific endeavour is in need of rehu-
manising even if it means a coming down from lofty but sterile heights to a more
mundane level. It is hoped that the final result will please those who nostalgically
think back to the times when scientific journals abounded with passionate contro-
versies, and when printing space could be made available to the scientist who wanted
to explain how loud the bang had been at an attempt to distill tetramethyllead, complete with a vivid description of how black the face of his assistant had been after the
explosion.*

If there is any major conclusion to be drawn from this work it must be that
Van der Waals effects are large and strongly variable with solvent, solute and even
with nuclear site within the solute molecule. The popularly held opinion that by
judiciously chosen referencing techniques the Van der Waals effects can be largely
eliminated must now be considered as debunked wishful thinking.

The writing of this review started three years ago; it would never have been finished but for substantial external help. We acknowledge with pleasure the University
of Regina for awarding a sabbatical leave, the National Research Council of Canada
and the Government of France for awarding a France-Canada Exchange Grant and
last, but not least, Mlle. Josien and Mme. Lumbroso-Bader of the CNRS Laboratoire
de Chimie-Physique for their hospitality in providing me with a quiet haven, where
most of this review was written. The experience has left me in awe for those who
manage to write books without such extensive privileges.

Chapter 1. Historical Development (Up to 1961)

The first mention of a medium effect for the chemical shift in high resolution NMR
was made in 1951 by W. C. Dickinson [1]. Based on classical magnetostatics he showed
that there is a finite contribution ΔH to the time averaged magnetic field at the
nucleus under study given by

$$\Delta H = \left[\frac{4\pi}{3} - \alpha \right] M \tag{1}$$

where $\frac{4\pi}{3} M$ is the Lorentz cavity field and $-\alpha M$ is the demagnetising field of the
bulk material surrounding the cavity. The factor α is a purely geometric factor; for a
sphere $\alpha = 4\pi/3$ (so that ΔH would be zero), but for a cylinder perpendicular to
the magnetic field $\alpha = 2\pi$, resulting in a non-zero ΔH:

$$\Delta H = \frac{-2\pi}{3} M = \frac{-2\pi\chi_v}{3} H_0 \tag{2}$$

---

* historical; search "Berichte".

where χ_v is the (volume) bulk susceptibility and H_0 is the externally applied field.

Combining Eq. (2) with the general expression $H \equiv H_0 + \Delta H = (1 - \sigma) H_0$, one then arrives at the well-known expression for the screening due to the bulk susceptibility;

$$\sigma_b = \frac{2\,\pi}{3} \chi_v = \frac{2\,\pi\chi_M}{3\,V} = \frac{2\,\pi\chi_M}{3\,M}\rho \tag{3}$$

where V is the molar volume, χ_M is the molar susceptibility and ρ is the density (Note; for a cylinder *parallel* to the magnetic field the geometric factor is $\alpha = 0$ so that $\sigma_b = \dfrac{-4\,\pi}{3} \chi_v$. This is of relevance for measurements with superconducting magnet systems).

Dickinson used this term to correct his experimental data. He also checked the validity of this correction, using paramagnetic solutions both in cylindrical and spherical cells. While he did find that the $2\,\pi\chi_v/3$ term was correct, provided that the length-to-diameter ratio of the sample tube was at least 10 : 1 or 20 : 1, he also found clear evidence that there was another shielding term present as well which had to be intermolecular in nature. Dickinson, after rejecting such possible causes as residual average of the fluctuating magnetic dipolar fields and nuclear dipole orientational effects as entirely negligible, found the reasons for these extra shifts in the paramagnetic nature of his solutions. Such effects indeed are very strong and usually obscure by far the much weaker other effects such as σ_w.

Although the σ_b term does not represent a real medium effect in the molecular sense of the word it is still necessary to discuss this term in the present review mainly for two reasons. Firstly, as we shall see, subsequent work on medium effects was initially discussed in terms of "deviations from the classical magnetostatic behaviour." Beyond this historical reason there is a very mundane second reason, since in fundamental studies of medium effects it is mandatory to use an external standard in order to exclude any medium effects on the reference compound (see Chapter 15). While it is easy to use a sample tube consisting of two coaxial cylinders, no one has ever succeeded in making a sample system consisting of two concentric spheres (for which the σ_b terms would be zero)!

The first attempts to a systematic investigation of medium effects were made by Bothner-By and Glick. In their first note [2] they point out that σ_b cannot be the only medium term since the aromatic protons of t-butylbenzene shift about 0.5 ppm less than the methyl protons upon 10-fold dilution with CCl_4. However, they tentatively ascribe all such extra effects to preferred orientation of the magnetically anisotropic neighbouring molecules. In their second paper [3] they report on binary liquid systems over the entire 0–100% concentration range, all relative to an external standard (H_2O) in a coaxial system. If $\Delta\chi$ is the difference in susceptibility of "solute" and "solvent" then, if σ_b is the only medium effect, all resonance lines should undergo a (linear) displacement of $2\,\pi\Delta\chi/3$ in going from the 0% to the 100% end of the concentration scale. For all their systems they found such a linear displacement, but rather than a constant $2\,\pi/3 = 2.09$ for all systems, a proportionality constant varying from 2.30 to 3.00 (for an average of 2.60) was found.

8

In Fig. 1 the data of Bothner-By and Glick are given in a graphical fashion. This graph, better than tabulated proportionality constants, shows the excellent linear relation between the dilution shifts and $\Delta\chi$, as well as the small degree of deviation of the individual points. The authors also proved, by measuring a few systems both in cylindrical and in spherical sample tubes, that the shape factor of $2\pi/3$ was correct. They additionally showed that the effect was a genuine screening phenomenon; changing the major field strength did not change the proportionality constant. They also rejected several other possibilities; the fact that the molecule may reside in a cavity that is ellipsoidal rather than spherical, and the subsequent change in cavity polarisation was found to be negligible. Equally unimportant is the direct magnetic polarisation of neighbouring molecules due to the induced magnetic moment of the

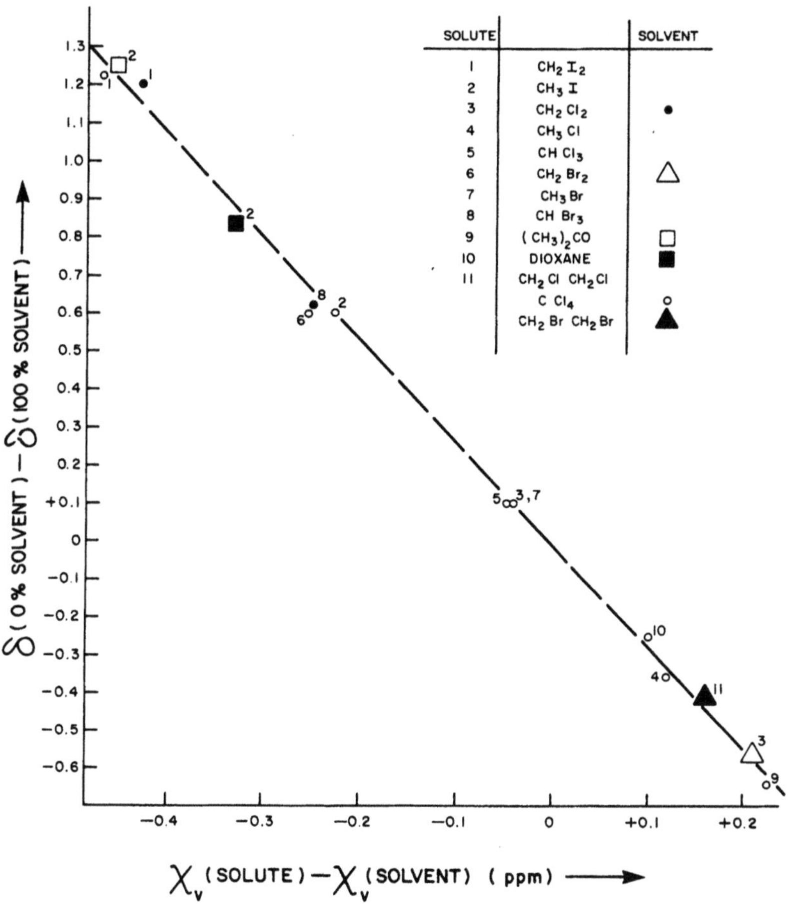

Fig. 1. Dilution shifts plotted against susceptibility difference. Average slope −2.60. Data from Bothner-By and Glick [3]

molecule under study (it should perhaps be mentioned here that Bothner-By and Glick again found that aromatic solvents behaved quite differently [4], which finding constitutes the origin of the study of the neighbour anisotropy effect σ_a, a topic which is, however, outside the scope of this review).

Glick and Ehrenson [5], following a suggestion of Bothner-By and Glick [3] used even the excellent linear relation between dilution shifts and $\Delta\chi$ to determine molar susceptibilities. However, Glick, Kates and Ehrenson [6] also showed that actually this proportionality constant is only a good constant (± 0.02 ppm) for any one given solute in any series of solvents, but that the constants for a number of solutes (each measured in the same series of solvents) have a range of about 25%. In fact they established the existence of a strictly linear relation between these constants and the inverse molar volumes of the solutes.

One can well imagine the puzzlement the early investigaters must have felt; on one hand the correctness of the bulk-susceptibility factor σ_b with its constant $2\,\pi/3 \approx 2.094$ was firmly established. But on the other hand there were the striking experimental facts on the dilution shifts requiring an extra contribution, but one that was equally proportional to $\Delta\chi$. Two distinctly different steps were required to solve this apparent contradiction. Firstly it required the notion that the extra effect is due to Van der Waals forces, as was subsequently guessed by Bothner-By [7] and secondly the realisation that the Van der Waals effects are roughly proportional to the diamagnetic susceptibility as derived later by Linder [8].

In the last paper of his series [7], Bothner-By succeeded in putting the phenomenon of medium effects on shielding on a sound footing. Rather than looking at *changes* in shift with concentration in binary systems, the medium effect for each individual solute was determined by measuring that solute first in the gas phase and then at infinite dilution in a solvent. (All compared to the same external reference). The difference between these two is then the pure gas-to-liquid or gas-to-solvent medium shift of that solute. If the gas measurement was at low pressure then the only way the susceptibility enters is in the measurement of the liquid and then it is the bulk susceptibility of the solvent only. Therefore

$$\sigma_l - \sigma_g = \sigma_b \text{ (solvent)} + \sigma \text{ (inter)} \tag{4}$$

Bothner-By found that for non-aromatic solvents the "excess shift" ($\sigma_l - \sigma_g - \sigma_b$) was negative. In other words, even after correction for σ_b there remained a downfield shift. Thanks to improved accuracy it also became clear that this factor could no longer be accounted for by a constant fraction of σ_b. The remaining term was identified as an intermolecular interaction and therefore a function of solute and solvent. It was also found that each solute could be assigned a characteristic number x_i and each solvent a characteristic number y_j so that the product $x_i y_j$ was equal (within experimental error) to the observed "excess shift" of solute i in the solvent j. This observation, incidentally, is the forerunner to the application of factor analysis (see Chapter 12) to medium shift effects.

In Table I we reproduce Bothner-By's results for non-polar solutes in non-polar isotropic solvents (i.e. where "excess shift" can be identified with σ_w).

Table 1. Excess shifts (= σ_w) gas-to-liquid (ppm) according to Bothner-By [7]. Data recalculated using better susceptibility Data[22]

Solvent Solute	Neopentane	Cyclohexane	Cyclopentane	Carbon tetra-chloride
C_5H_{10}	−0.18	−0.16	−0.17	−0.30
C_2H_6	–	−0.19	−0.18	–
$C(CH_3)_4$	−0.21	−0.19	−0.20	−0.34
C_2H_4	−0.23	−0.21	−0.22	–
$Si(CH_3)_4$	–	−0.24	–	−0.37

After rejecting any orientational effects for such systems as indicated in Table 1, Bothner-By then suggested dispersion forces as the probable origin.

In fact this suggestion, as duly acknowledged by Bothner-By, had been made nearly two years previous. Marshall and Pople [9] in 1958 published a study of the effect of electric fields on the shielding of an H atom. For this spherically symmetric ($1s$) electronic configuration they found that a static electric field E, either parallel or perpendicular to the magnetic field will cause a deshielding proportional to E^2. The effect is caused both by a reduction of the Lamb term (decrease of the diamagnetic term) as well as (in the perpendicular case) by a partial upset of the otherwise perfect circular rotation of the electronic system around the magnetic field axis (the so-called "paramagnetic" term). The shielding of a $1s$ hydrogen atom in an electric field is then given as

$$\sigma_{||} = \frac{e^2}{3\,mc^2 a} \left[1 - \frac{439}{40} \frac{a^4 E^2}{e^2} \right] \tag{5}$$

$$\sigma_{\perp} = \frac{e^2}{3\,mc^2 a} \left[1 - \frac{193}{15} \frac{a^4 E^2}{e^2} \right] \tag{6}$$

where a is the Bohr radius $\hbar^2/me^2$

M. J. Stephen [10], in the same issue of Molecular Physics as where Marshall and Pople's paper appeared, argued that for all non-polar isotropic molecules the intermolecular shielding should be

$$\sigma_w = \left(\frac{\Delta\sigma_{||} + 2\,\Delta\sigma_{\perp}}{3} \right) \overline{F^2} = -B\overline{F^2} \tag{7}$$

where $\overline{F^2}$ is the non-zero average of the square of the instantaneous electric fields generated even by non-polar molecules and $\Delta\sigma_{||}$ and $\Delta\sigma_{\perp}$ are the changes in shielding of the "solute" molecule due to $\overline{F^2}$. The averaging of $\Delta\sigma_{||}$ and $\Delta\sigma_{\perp}$ as indicated in Eq. (7) assumes absence of preferred orientation between solute and solvent. However, Stephen also advanced two reasons why this dispersion shielding term would be too small to be detected, at least for CH_4. He pointed to an estimate of $\overline{F^2} = 1 \cdot 10^{11}$ esu for liquid methane made earlier by Buckingham and Stephen [11] based on a combination of classical dielectric theory plus the heat of evaporisation of liquid methane. Then, assuming a minimum value of $(\Delta\sigma_{||} + 2\,\Delta\sigma_{\perp})/3 = 0.73 \cdot 10^{-18}$ esu (based on the calculations for the 1s hydrogen atom of Marshall and Pople, Eqs. (5) and (6), the dispersion shielding as per Eq. (7) would be equal to $0.73 \cdot 10^{-7}$. At 40 MHz this is equivalent to 3 Hz, an effect indeed on the border of detectability in those days.

Stephen then points to a 1958 study by Schneider, Bernstein and Pople [12] and states that these investigators indeed had found a zero gas-to-liquid shift for methane. Careful reading of reference [12] indicates, however, that the authors gave the zero shift only as a rough estimate, not as a measured value, the reason being that both the density (at -98 °C) and the molar susceptibility of CH_4 were not known to the authors, so that no σ_b correction could be applied to the experimental gas-to-liquid shift.

This misunderstanding probably contributed to a considerable delay in the development of the theory for σ_w. This is the more remarkable since Schneider, Bernstein and Pople [12] could have given a reasonable estimate of σ_b for CH_4. Densities for CH_4 at $-160, -150, -140$ and -82.5 °C were known at that time [13] from which a maximum value of $\rho = 0.33$ (at -98 °C, the temperature of the actual shift measurement) can be estimated. Together with a χ_M value estimated from Pascal's constants a maximum value for σ_b at -98 °C of 0.76 ppm is then obtained. Compared with the experimental, uncorrected, gas-to-liquid shift of 1.06 ppm which can be extracted from Schneider, Bernstein und Pople's paper [12], the evidence of an "excess" shift appears to have been there for the taking. In the same paper, the gas-to-liquid shift for ethane is also indicated as being zero. In this case a σ_b correction was indeed made, but ironically enough the authors used obsolete density data, while better values were available at that time [13] so that again the σ_w effect accidentally escaped detection. As a matter of record it might be proper to note that Schneider, Bernstein and Pople [12] did not discuss the zero magnitude of the gas-to-liquid shifts of methane and ethane *at all*. The main importance of their work is the establishment of sound experimental procedures (although without quite the required accuracy) and more to the point, the production of the first study on the influence of hydrogen bonding on medium shifts.

Fortunately enough another 1958 communication by Evans [14] did report a 0.50 ppm "excess" downfield shift for CH_4 gas-to-infinite-dilution-in-CCl_4. Evans' paper is also the first to report on ^{19}F gas-to-liquid shift. His reported value of 7.77 ppm of "excess" shift of CF_4 gas-to-infinite-dilution-in-CCl_4 is a very important finding. Not only is the ^{19}F σ_w effect apparently much larger than for protons (and therefore more easily detectable) but this large magnitude also helps considerably in finding the origin of the effect. Evans also reported the ^{19}F resonance frequency of trifluoro-benzene in infinite dilution in various solvents as compared to the neat liquid. The results span a range of 8 ppm, a magnitude that in no way could be accounted for by the difference in bulk susceptibility of the solvents, a very clear indication of the existence of another superimposed effect.

We can now return to Bothner-By's 1960 paper [7]. We have already discussed some of his results and we just finished recounting the theoretical and experimental basis available to him. Equating London's expression for the interaction energy due to the fluctuating field F of the (non-polar) solvent molecules with the classical energy of a polarisable solute molecule in an electric field he finds

$$\overline{F^2} \approx \frac{3\,h\nu\,\alpha_2}{2\,r^6} \tag{8}$$

where $h\nu$ is the mean excitation energy, α_2 the polarisability of the solvent and r the intermolecular distance between solute and solvent molecule. Then, taking $h\nu \approx 50.000$ cm^{-1}, $r \approx 5$ Å, $\alpha \approx 50$ Å^3 and $\sigma \approx -10^{-18}\,\overline{F^2}$ (from Marshall and Pople [9]), Bothner-By finds that Eqs. (7) and (8) lead to a σ_w of about -0.1 ppm, which is indeed the order of magnitude he observed for non-polar systems. No attempt was made to use Eq. (8) for the calculation of the individual gas-to-liquid shifts and perhaps one can surmise why. First of all, as was later shown by Raynes, Buckingham and Bernstein [15] a more appropriate expression for F^2 is:

$$\overline{F^2}\,(\text{pair}) = \frac{3\,\alpha_2 I_2}{r^6} \tag{9}$$

where I_2 is the ionisation potential of the solvent. Eq. (9) gives results 2 to 3 times larger than those calculated using Eq. (8). Furthermore, there is the problem of determining the intermolecular distance r. (It is most interesting to note that Bothner-By in this connection talks about the *accessibility* of the solute protons to the field of the solvent molecules; in fact he predicts that different protons in a solute molecule will have different solvent effect due to this. Such a site factor was indeed shown to exist, but not until several years later by Rummens and Bernstein [16, 17]). Yet another formidable problem is that of expanding models for interacting pairs [such as Eqs. (8) and (9)] into a more realistic model of one solute molecule surrounded by many solvent molecules at an infinite array of distances.

At about the same time as Bothner-By [7], Buckingham, Schaefer and Schneider published an extensive study of the gas-to-solution shift of CH_4 in a variety of solvents [18]. We reproduce in Table 2 their results for CH_4 in non-polar isotropic solvents. (Unless indicated otherwise, we have re-calculated in all Tables the σ_b term — and hence the σ_w term — using up-to-date data for χ_M [19–22]).

In an attempt to obtain some information on the nature of the σ_w, the authors plotted σ_w versus H_b, the heat of vaporisation at the boiling point (obtained from) the empirical Hildebrand-Scott relation $H_b = 17.0\,T_b + 0.009\,T_b^2$). It was found that the points for the hydrocarbon solvents of Table 2 are on a straight line, with a proper (*i.e.* negative) slope. Interestingly, the points for polar solvents like *cis*-butene-2, acetone, ethylether, ethylacetate, ethylnitrate and triethylamine fell on the same line. The line does not go through the origin, however, (intercept at +0.25 ppm), while in addition the halogen-containing solvents of Table 2 lie on a different straight line, parallel to the first one (and, incidentally, going through the origin).

Attempts to correlate σ_w to other properties that are also determined by Van der Waals dispersion forces are a natural thing to do, but, as the relative failure of the just quoted work indicates, there are a number of pitfalls.

Table 2. $-\sigma_W$ (ppm) for CH_4 (gas-to-5% solution) according to
Buckingham, Schaefer and Schneider [18]
Note; gas measurements were actually done with 10 atm CH_4;
to obtain shifts relative to zero atm CH_4, 0.025 ppm has been
added to all "experimental" data. The σ_b correction has been
adjusted, but assuming infinite solute dilution

Solvent	$-\sigma_W$ (ppm)
neopentane	0.24
cyclopentane	0.27
n-hexane	0.26
cyclohexane	0.30
trans-butene-2	0.16
CCl_4	0.44
$SiCl_4$	0.29
$SnCl_4$	0.39

In 1961 Gordon and Dailey [23] published an important study on methane,
ethane and ethylene both in the vapour and liquid state. The authors had the considerable advantage of having available the just then published accurate χ_M data for
the molecules studied [19]. The main results are given in Table 3.

Table 3. $-\sigma_W$ (ppm) of CH_4, C_2H_6, C_2H_4 according to Gordon and
Dailey [23]

	$-\sigma_W$ (gas)	$-\sigma_W$ (liquid)
CH_4	$0.482\,\rho$	0.176 experimental at $t = -106.1\,°C$ 0.156 extrapolated from gas data
C_2H_4	$0.492\,\rho$*)	$0.488\,\rho$
C_2H_6	$0.390\,\rho$	$0.515\,\rho$

*) The authors did not give this number, but it can be extracted
   from other data they provided.

The authors state that the density dependence of σ_W is "similar" for the gaseous
and the liquid states (although they also indicate the difference for ethylene is about
3 x their estimated point accuracy of ±0.025 ppm). The paper is somewhat lacking
because the authors did not state at what temperature most of the measurements
were done nor did they give a reference to the liquid densities they used. Nevertheless,
the idea of measuring compounds in both the liquid and the gaseous state is an important one. Not until ten years later was their main conclusion — essentially the
same linear relation between σ_W and density ρ throughout gas and liquid state —
challenged in a study on ethane with an order of magnitude better accuracy [24].

Gordon and Dailey also had to make the assumption that the shielding of CH_4
had a zero *intra* molecular temperature dependence. This important matter will be
taken up in Chapter 11, which deals with the temperature dependence of σ_W.

At this arbitrarily chosen point we wish to discontinue the more or less chronological order. Apart from a few publications which will be discussed later the coverage to this point has been comprehensive and complete to our knowledge. To maintain the chronological and comprehensive structure after 1961 would result in a fragmentation that is one of the objectives of a review to avoid.

Chapter 2. Continuum Models

2.1. The Continuum Model of Linder

In 1960 Linder [8] published a continuum model for the calculation of non-polar interaction energies.

Analogous to the Onsager theory for the reaction field R of permanent dipoles in a continuum, Linder developed a model based on the reaction field R^* of a spontaneous oscillating dipole. Suppose the central molecule has a moment $m(v_i)$. At a distance r_k this causes a fluctuating dipole field $F_k(v_i)$. This field induces a moment $m_k(v_i)$ in each molecule of the continuum.

$$m_k(v_i) = \alpha_k^* F_k(v_i) \tag{10}$$

It should be noted that α_k^* is different from the static polarisability α_k, because the oscillator frequency of the k molecules may be different from that of the i molecule (if the molecules i and k are of the same species the average frequencies are equal, but at any given moment v_i is different from v_k). One has therefore

$$\alpha_k^* = \alpha_k \; \frac{v_k^2}{v_k^2 - v_i^2} \simeq \alpha_k \; \frac{v_j^2}{v_j^2 - v_i^2} \tag{11}$$

where it has been temporarily assumed that all molecules k have the same frequency v_j. Each moment m_k causes a reaction field contribution, which, added together, form the reaction field R^*

$$R^* = g_j m(v_i) \; \frac{v_j^2}{v_j^2 - v_i^2} \tag{12}$$

Eq. (12) is analogous to the expression for the reaction field $R = gm$ of a permanent induced moment m, the difference being the $v_j^2/(v_j^2 - v_i^2)$ term. The latter, after integration and averaging, is going to cause a factor of 1/4 difference in the work expression. The reaction field factor is given by

$$g_i = \frac{2n^2 - 2}{2n^2 + 1} \; \frac{1}{a_i^3} \tag{13}$$

Linder then evaluates the work function W (*i.e.* the work done to bring the cavity into the dielectric)

$$W = -1/2m(v_i) \cdot R^* = -1/2\overline{m(v_i)^2}g_j v_j^2/(v_j^2 - v_i^2) \tag{14}$$

where already the averaging over all possible values of $m(v_i)^2$ has been carried out. The remaining problem is the integration over the frequency distributions $\rho(v_i)$ and $\rho(v_j)$ respectively. This is facilitated by first introducing (assuming a quantal oscillator)

$$\overline{m(v_i)^2} = \frac{3}{2}h v_i \alpha_i \tag{15}$$

After substitution of Eq. (15) into the integrated form of Eq. (14), using the approximate equality $\alpha_i g_j = \alpha_j g_i$ and using the assumption that the average values for v_i and v_j are the same and equal to the natural frequency v_0 of a single oscillator one finds

$$W = -\frac{3}{16}h v_0 \alpha_g = -\frac{1}{8}\overline{m_0^2}g \tag{16}$$

The difference between Eq. (16) and the analogous expression $W = -mR/2 = -gm^2/2$ for the reaction field due to a *permanent* moment should be noted.

For mixtures of two distinct species the derivation of W is similar except that the average values of v_i and v_j are now different. From a more generalised treatment of the model, also due to Linder [25], one finds

$$W = -1/4 m_1^2 g \frac{v_2}{v_1 + v_2} \tag{17}$$

where the indices 1 and 2 refer to solute and solvent respectively. Eq. (17) can be combined with the expression for the potential of a polarisable molecule in an electric field. For a static field this would be equal to $-\alpha_1 E^2/2$. For a fluctuating field this relation still holds, according to Linder, provided the mean square field $\overline{F^2}$ is used. One has therefore

$$\overline{F^2} = \frac{m_1^2 g}{2\alpha_1} \frac{v_2}{(v_1 + v_2)} \tag{18}$$

Eq. (18) combined with $\sigma_w = -B\,\overline{F^2}$ [Eq. (7)] then gives the resulting expression for σ_w. One remaining problem is that of evaluating m_1^2. Linder indicates several possibilities, the most logical one being the expression for a ground state quantal harmonic oscillator [Eq. (15)], which then leads to

$$\sigma_w = \frac{-3 B\, hg\, v_1 v_2}{4(v_1 + v_2)}$$

$$= -\frac{3 B (n_2^2 - 1)\, h\, v_1 v_2}{2(2 n_2^2 + 1)\, a_1^3 (v_1 + v_2)} \tag{19}$$

The energies $h\nu_i$ are really the transition electronic energies ΔE_i from the ground state to the first excited, non-forbidden state. Although ΔE_i is always considerably less than the ionisation potential I, one nevertheless uses often the London approximation

$$\nu_i = I_i/h \tag{20}$$

In combination with Eq. (19) this should cause a calculated σ_w which is systematically somewhat too high.

Another possibility, also given by Linder, is to combine Eq. (15) with

$$m_i^2 = \frac{-6\,m_e c^2}{N}\chi_M \tag{21}$$

where χ_M is the molar diamagnetic susceptibility.

This leads to

$$\nu_i = \frac{-4\,m_e c^2}{h\alpha_i N}\chi_{M_i} \tag{22}$$

which can then again be combined with Eq. (19) to give an expression for σ_w. It is in Eq. (22), combined with Eq. (19) that one notices the relation between medium shift and susceptibility which early investigators observed, as discussed in the previous Chapter. Also the inverse proportionality to the molar volume of the solute, as noted by Glick *et al.* [6] is present in Eq. (19), through the reaction field factor g [Eq. (13)].

Howard, Linder and Emerson [26] compared the $\overline{F^2}$ obtainable from Linder's theory to observed gas-to-liquid shifts. In Fig. 2 we have reproduced those data which refer to σ_w shifts and then only for the approximation of Eq. (20). The results are typical, however, not only for the Linder theory but perhaps for all continuum models. It appears that there is an almost linear relation between σ_w and $\overline{F^2}$, but one that is different for each solute, with slopes much larger than the theoretical prediction and, to make it all much worse, with distinctly non-zero intercepts. The authors also tried Eq. (22), but this gave no improvement.

In their calculations the authors used the Onsager approximation

$$a_i^3 = 3\,V_i/4\,\pi N \tag{23}$$

but this cannot be the reason of the discrepancies either. In Chapter 5 we will return to the Linder model in a discussion of the physical meaning of the field $\overline{F^2}$.

Subsequently, Lumbroso, Wu and Dailey [27] made another study of Linder's continuum model. Values for σ_w were measured for CH_4, C_2H_6, C_2H_4 and cyclopentane in cyclohexane, dioxane and CCl_4. Using both the Onsager approximation [Eq. (23)] and the London approximation [Eq. (21)] they again found that the predicted σ_w values were to low by a factor 2 to 3 with a poor correlation between $\overline{F^2}$ and observed σ_w shifts. The relevant experimental data are given in Table 8. (Chapter 4)

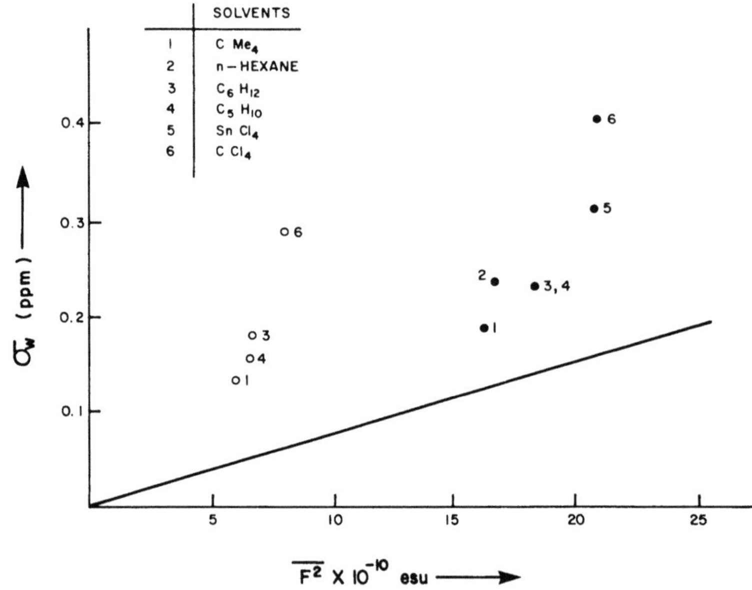

Fig. 2. Van der Waals shifts σ_W plotted against the average squared electric field $\overline{F^2}$ of Eqs. (18) (15) and (20). Open circles refer to solute cyclopentane (σ_W data from Bothner-By [7], Table 1); filled circles for CH_4 as solute (σ_W data from Buckingham et al. [18], Table 2). After Howard, Linder and Emerson [26]. Solid line refers to theoretical relation obtained by Marshall and Pople [9] as given in Eq. (7)

The reasons for the apparent failure of Linder's model may be manyfold. The very concept of the Onsager model — a cavity in a continuous medium — is not realistic. It may be argued that this same model has met considerable success in the theory of dielectrics such as the calculation of dipole moments from dielectric constants. However, with Van der Waals forces one is dealing with an r^{-6} dependence (rather than the r^{-3} dependence of dipolar fields), which restricts the interaction pretty well to the first "shell" of molecules around the solute. It should be no wonder then, that the discontinuity at the cavity-dielectric interface as well as the insistence that the first solvent shell has identical properties as the bulk solvent, are particularly inappropriate for calculations of Van der Waals effects. It may be remarked here that Linder's model also allows the calculation of the (free) energy of cohesion of non-polar liquids [8, 28]. In comparison to experimental heats of vaporisation these calculated cohesion energies are much too low for "normal" organic liquids (C_6H_{12}, C_6H_6, CCl_4), about twice too high for liquid CH_4 and even 4 to 5 times too high for smaller molecules like H_2 or He. It follows therefore, that the discrepancies noted previously cannot be exclusively due to the magnetic aspects of Eq. (19).

Other approximations such as Eqs. (15, 20, 22 and 23) as well as the inherent assumption of *one* resonance frequency only may cause perhaps as much error as the Onsager model itself. In section 5.3 the physical meaning of Linder's effective electric field will be further discussed.

In addition, the neglect of the site factor (see Chapter 6) is certainly a major contributing factor. Fortunately this factor can be calculated with good accuracy (except for large or strongly non-spherical molecules as discussed in section 6.1). Division of experimental σ_w data by calculated site factors would yield parameters, more closely related to Linder's theory. Note, however, that such a procedure would be rather artificial, since in any true continuum model the reaction field vector is constant over the entire cavity. There is thus no suitable basis for the notion of a site factor *at al.*

2.2. The Continuum Model of De Montgolfier

Since about 1966 a group of French scientists and in particular Ph. de Montgolfier have attempted an alternate approach to the continuum model for the calculation of medium shifts. In the first paper of this series, Barriol and De Montgolfier [29] start with the following relation for the interaction energy W in a non-polar liquid

$$W = -\frac{1}{2} g \, \overline{m^2} \tag{24}$$

which, coupled with the expression for $\overline{m^2}$

$$\overline{m^2} = \frac{3}{2} \alpha \, \Delta E \tag{25}$$

(with ΔE being the unique transition energy of the oscillator), leads to

$$W = -\frac{3}{4} g\alpha \, \Delta E \tag{26}$$

Upon comparing Eq. (26) with Linder's corresponding expression for W [Eq. (16)] one notices that Eq. (26) gives a four times higher energy. Since Eq. (25) is identical to Eq. (15), the difference must have been caused by Eq. (24). Barriol and De Montgolfier did not *derive* this equation but transposed it directly from the theory of dielectric polarisation with regard to *permanent* non-polarisable dipoles. As the careful, step by step derivation of Linder has shown [see Eq. (16) and preceding discussion], however, such a transposition is not valid.

Barriol and De Montgolfier do divide all their results by a factor of 2, however, on the basis of the fact that Eq. (26) gives numerical results twice those of the corresponding London equation. This still leaves an extra factor of 2 in comparison to Linder's model. Since, as indicated before, Linder's model often calculates energies or medium shifts that are considerably too small, De Montgolfier's theory has the gratifying, if accidental, effect of being more accurate.

Using the Onsager relation [Eq. (23)] and the Lorentz-Lorenz equation

$$\frac{n^2 - 1}{n^2 + 2} V = \frac{4}{3} \pi N \alpha \tag{27}$$

to eliminate α, Barriol and De Montgolfier then arrive at

$$W = -\frac{3}{4} \frac{(n^2 - 1)^2}{(2n^2 + 1)(n^2 + 2)} \Delta E \tag{28}$$

[included in Eq. (28) is the division by 2].

By equating W to the experimental heat of vaporisation of a number of neat liquids (CCl_4, C_6H_6, n-C_5H_{12}, C_6H_{12}) the authors then find very reasonable values for ΔE (or rather for the corresponding wavelengths of absorption) for these liquids.

In a subsequent paper, De Montgolfier [30] uses Eq. (28) to arrive at

$$\sigma_w = -\frac{kB\, 6(n^2 - 1)^4}{(2n^2 + 1)^2 (n^2 + 2)^2} \left(\frac{\Delta E}{\alpha} \right) = -6 \left(\frac{kB\Delta E}{\alpha} \right)_1 f(n) \tag{29}$$

where n is the refractive index of the solvent and k is a geometric factor which we will discuss later. A plot of σ_w versus the refractive index function should therefore yield a straight line through the origin. De Montgolfier uses for this plot again the gas-to-solution shifts of CH_4 in a variety of non-polar (and polar) solvents as determined by Buckingham, Schaefer and Schneider [18]. A remarkably good straight line was obtained which passes indeed through the origin. Furthermore, from the slope a value of $k = 2.5$ is obtained in exact agreement with previous calculations of De Montgolfier on the k factor of CH_4 [31]. One should note, however, that certain solvents, notably CH_2Br_2, $CHBr_3$, CH_3I and CH_2I_2 were left out (these points do not fall on the straight line!). The otherwise excellent agreement is furthermore marred by the fact that in the derivation of Eq. (29) several errors were made. To see this, the derivation will be reproduced below, with one important difference; we will keep track of what is a solute property and what is a solvent property by assigning subscripts 1 or 2 respectively to all molecular parameters. De Montgolfier starts with the observation that the reaction field R of a permanent moment μ can be written as $R = g \cdot \mu$. It is then stated that for *instantaneous* dipole moments m then has by correspondence

$$\overline{R^2} = g^2\, \overline{m_1^2} \tag{30}$$

As said before it seems doubtful, however, that such a direct transposition is valid [see also Eq. (12)]. Then, using Eq. (25), it follows that

$$\overline{R^2} = \frac{3}{2} (g\,\alpha_1)^2 \frac{\Delta E_1}{\alpha_1} \tag{31}$$

For non-polar liquids, for which $\epsilon = n^2$, one can combine the expression for the reaction field factor [Eq. (13) with n_2^2 and cavity radius a_1] with the Lorentz-Lorenz equation [Eq. (27) with n_1, V_1 and α_1] to arrive at

$$\overline{R^2} = \frac{6(n_2^2 - 1)^2 (n_1^2 - 1)^2}{(2n_2^2 + 1)^2 (n_1^2 + 2)^2} \frac{\Delta E_1}{\alpha_1} \tag{32}$$

It is already clear that for gas-to-infinite-dilution data Eq. (32) must be used, and not Eq. (29) which is intended for neat liquids only. It might be mentioned here that De Montgolfier in his derivation refers to some earlier work of Barriol and Weissbecker [32], in which paper exactly the same error is made.

De Montgolfier then further continues by writing

$$\overline{F^2} = k \, \overline{R^2} \tag{33}$$

and

$$\sigma = -B \, \overline{F^2} = -kB \, \overline{R^2} \tag{34}$$

when then results in

$$\sigma = \frac{-6 \, kB (n_2^2 - 1)^2 (n_1^2 - 1)^2}{(2 \, n_2^2 + 1)^2 (n_1^2 + 2)^2} \frac{\Delta E_1}{\alpha_1}$$

$$= \frac{-8 \, \pi \, kB \, N (n_2^2 - 1)^2 (n_1^2 - 1)}{3 (2 \, n_2^2 + 1)^2 (n_1^2 + 2)} \frac{\Delta E_1}{V_1} \tag{35}$$

The factor k, which is usually greater than 1, results from De Montgolfier's consideration that the reaction field is not homogeneous, due to the anisotropy of the bond polarisabilities of the solute molecule inside the cavity. For further discussion of k see section 6.3 dealing with this site factor. It may be noticed that the above derivation is void of any thermodynamic argument, which is achieved via Eq. (33). However, the two electric fields $\overline{F^2}$ in Eqs. (33) and (34) are not necessarily the same, as will be further discussed in section 5.4.

In a follow-up paper [*33*], De Montgolfier shows that the chemical shifts of C_5H_{10} or C_6H_{12} measured relative to *internal* TMS plotted versus the (solvent) refractive index function of Eq. (29) yields an extremely good straight line (± 0.1 Hz) for solvents like acyclic and cyclic saturated hydrocarbons and for CCl_4. In our view, however, such internally referenced shifts should be plotted against the difference of *two* refractive index functions as per Eq. (35), with the index 1 referring once to the solute and next to the TMS. It was also noticed that for solvents like acetone, diethylether and chloroform, the points are not on the line at all., although the dispersion theory does not distinguish between polar and non-polar solvents. The suggestion of De Montgolfier that these deviations are due to the magnetic anisotropy (σ_a) of these solvents does not appear too likely, since σ_a is only weakly influenced by the solute [*34*] so that both the solute (C_5H_{10} or C_6H_{12}) and the TMS will probably experience the same σ_a; so the two effects should almost cancel out. In addition, *externally* referenced measurements were done (no actual numerical data are given in this paper, however). For TMS as a solute in various solvents the chemical shifts relate reasonably well to the proposed refractive index function. In fact the point for gaseous TMS is right on the line (corresponding to fulfilment of the "through-the-origin" condition for gas-to-liquid shifts), but in addition to the polar solvents now also C_6H_{12} and methyl-cyclohexane $CH_3 - C_6H_{11}$ are off. In a similar plot for C_5H_{10} as a solute, the same is found except that now in addition the point for gaseous C_5H_{10} is 0.05 ppm off.

Since in the last two examples external referencing was employed, deviation of the straight line could well be interpreted in terms of σ_a effects. Inspection of the graphs given by De Montgolfier show, however, that one would have to assume for acetone a σ_a effect of $\sigma_a = -0.05$ ppm for TMS as a solute and $\sigma_a = +0.05$ ppm for C_5H_{10} as a solute. This seems altogether unacceptable.

Another curiosity is that De Montgolfier in none of his papers ever discussed CS_2 as a solvent; because of its very high refractive index it would provide a very good test for any continuum model. Furthermore medium shift data in CS_2 are available. From a few tests it has appeared to us that inclusion of CS_2 as a solvent would have resulted in major deviations in all cases.

De Montgolfier's work has been further tested by Chenon [35] and Chenon, Bouquant and Lumbroso-Bader [36]. Their experimental gas-to-liquid data are given in Table 4. In this table only non-polar isotropic solvents were included. The authors in fact plotted the externally referenced (to C_6H_{12} liquid) shifts versus De Montgolfier's function f(n) of Eq. (29) and attempted to draw a straight line through these points. They anchored this line by having it go through the CCl_4 point and through the points relating to the n-paraffins as solvent. They then found, as did De Montgolfier, that the gas points are off by 0.05 ppm (i.e. $n = 1$ gives 0.05 ppm too large a shielding for the gaseous n-hexane or for C_6H_{12}) while also the three cyclic solvents (see Table 4) are off in the same direction and by about the same amount.

Table 4. $-\sigma_w$ (ppm) at 41 °C according to Chenon *et al.* [35, 36]

Solute \ Solvent	n-hexane	n-pentane	n-heptane	n-octane	C_6H_{12}	CCl_4	1,4-dioxane	tetra hydro pyran
n-hexane (CH_3)	0.175	–	–	–	0.207	0.298	0.230	0.202
n-hexane (CH_2)	0.138	–	–	–	0.163	0.248	0.190	–
C_6H_{12}	0.162	0.155	0.173	0.182	0.192	0.276	0.203	0.182

However, it could be pointed out that if a best straight line was forced through the gas point, all data of Table 4 would fall within 0.03 ppm of that line.

The same authors also included n-heptane and a number of isomeric 1, 3, 5-trimethylcyclohexanes in their studies. No vapour measurements were done on these solutes, but as before an external reference was used. Data are given in Table 5. Plotting these data in the usual fashion gave again approximately linear correlations. Although in the latter case the intercept cannot readily be interpreted, the slopes could be extracted even though no gas point was available. Provided ΔE_1 and α_1 are known, the combined parameter kB of Eq. (29) could then be found.

For the data of Tables 4 and 5 this was done [36] as reproduced in Table 6.

It may be noted that these slopes vary some 20% although the solutes are really very similar in chemical nature. It appears likely that most of this variation is due to the neglect of the site factor. This is corroborated by the fact that without exception these slopes are larger for the *outer* protons as compared to those of the inner protons of the same solute.

Table 5. Shifts in ppm at 29 °C, upfield from external liquid C_6H_{12} (corrected for bulk susceptibility) according to Chenon *et al.* [35, 36]

Solvent Solute	*n*-pentane	*n*-hexane	*n*-heptane	*n*-octane	CCl_4
n-heptane ($C\underset{\sim}{H}_3$)	–	–	0.568	–	0.452
n-heptane ($C\underset{\sim}{H}_2$)	–	–	0.173	–	0.077
1,3,5-trimethyl- cyclohexanes;					
trans, $(C\underset{\sim}{H}_3)_{eq}$	0.627	0.628	–	0.608	0.510
trans, $(C\underset{\sim}{H}_3)_{ax}$	0.483	0.483	–	0.470	0.368
cis, $(C\underset{\sim}{H}_3)_{eq}$	0.588	0.592	–	0.573	0.475
C_6H_6 (at 41 °C)	–	−5.743	–	−5.758	−5.915

Table 6. Empirical values of kB for hydrocarbons, according to Chenon *et al.* [35, 36]

Solutes →	*n*-heptane		*n*-hexane		C_6H_{12}	1,3,5-trimethylcyclohexanes		
	$C\underset{\sim}{H}_3$	$C\underset{\sim}{H}_2$	$C\underset{\sim}{H}_3$	$C\underset{\sim}{H}_2$		*trans*, $(C\underset{\sim}{H}_3)$ eq.	*trans*, $(C\underset{\sim}{H}_3)$ ax.	*cis*, $(C\underset{\sim}{H}_3)$ eq.
$10^{-3} \cdot 6\, kB\, \Delta E_1/\alpha_1$ (Hz)	5.00	4.14	4.73	4.22	4.42	4.11	3.87	4.05
$kB \cdot 10^{18}$ (esu)	5.74	4.75	4.79	4.27	4.27			

The kB factors are rather larger than the value of 2.5 calculated for CH_4 [*31*]. De Montgolfier has argued that this factor should be about the same for all saturated hydrocarbons, but he finds experimentally $kB = 2.5$ for CH_4 and $kB = 10$ for cyclopentane [*30*]. The values for kB of Table 6 are intermediate to these. It therefore seems that De Montgolfier's model does not have universal applicability. Even for solutes of comparable chemical nature rather different proportionality constants are apparently required. One may wonder, of course, to what extent this is related to the aforementioned errors in this model.

2.3. The McRae Formula

Many other modifications of the continuum model could be tried out on chemical shift data, but in our opinion it is not likely that anything noticeably better than the models of Linder and of De Montgolfier can be expected. It appears that one may select just about any function of dielectric constant or refractive index which, when plotted against solvent shift will give a straight line for one solute in a narrow range of solvents, while at the same time no single formula of universal applicability will emerge. A good example of this is Chenon's *et al.* attempt [*36*] to correlate σ_w's with McRae's function $(n_2^2 - 1)/(2\,n_2^2 + 1)$ [*37*]. They report that an equally good straight line is obtained as with De Montgolfier's function. They abandon McRae's formula

only because the gas points are off even more conspicuously (0.125 ppm). Also
Laszlo, Speert and Raynes [38] have tried a form of the McRae formula as follows

$$\sigma_m = C \frac{n_2^2 - 1}{2\,n_2^2 + 1} + D \tag{36}$$

For the gas-to-liquid shift of CH_4 in a number of halocarbon solvents they found an
excellent linear relation between σ_m and $f(n_2)$ (correlation coefficient 0.98), although
the line did not go through the origin (equivalent to the off-the-line position of the
gas point as found by Chenon *et al.*) and hence the requirement of a second constant
like the parameter D in Eq. (36). We feel, however, that these anomalies in both cases
may be due to a misunderstanding of McRae's formula; the $f(n_2)$ given above is
proportional to the reaction field itself and should therefore be squared before it is
compared to σ_w. Note that in electronic spectroscopy the gas-to-liquid shifts $\Delta\nu$ are
directly proportional to the reaction field as was shown by Bayliss [39]. We have
carried out this quadrature; it was found that indeed the linear correlation between
σ_w and $f^2(n_2)$ is just as good if not better than with Eq. (36) and that the gas point
is then almost right on this line. The same effect can be achieved by taking the square
root of De Montgolfier's $f(n)$ of Eq. (29). Note that in both cases one ends up with
a solvent dependence as prescribed by Eq. (35).

Chapter 3. Pair Interaction Models for σ_w

3.1. The Binary Collision Gas Model of Raynes, Buckingham and Bernstein

In 1962, Raynes, Buckingham and Bernstein [15] published an entirely different
theory for the calculation of medium shifts, which is basically a statistical mechanical model. All possible medium effects can be calculated in this theory, but in the
derivation given below we will restrict ourselves to σ_w. Using a general virial theorem
[40] one may write:

$$\sigma = \sigma_0 + \frac{\sigma_1}{V_M} + \frac{\sigma_2}{V_M^2} + \dots \tag{37}$$

where σ is the shielding in a medium, σ_0 the corresponding shielding of the solute
in vacuo, V_M is the molar volume of the medium (to be averaged on a mole fraction
basis in case of mixtures), σ_1 is the 2nd virial coefficient of shielding corresponding
to *binary* collisions while $\sigma_2, \sigma_3 \dots$ correspond to ternary and higher order collisions.
The authors showed [15] that for gases of medium pressure (up to 50 atm), the
shielding was strictly proportional to $1/V_M$, so that all but the binary collisions
could be ignored leading to;

$$\sigma = \sigma_0 + \sigma_1/V_M \tag{38}$$

The second virial coefficient still contains all kinds of contributions such as the bulk susceptibility term

$$(\sigma_1)_b = \frac{2\pi}{3} \chi_M \tag{39}$$

and the Van der Waals term, which takes the following form;

$$(\sigma_1)_w = \int (\sigma_w)_{pair} \exp(-u/kT)\, d\tau \tag{40}$$

where $d\tau$ is the coordination space element

$$d\tau = r^2 dr \sin\theta_1\, d\theta_1 \sin\theta_2\, d\theta_2\, d\phi \tag{41}$$

For the energy term in the Boltzmann factor the Lennard-Jones (6–12) potential was selected

$$u = 4\,\epsilon\left[\left(\frac{r_0}{r}\right)^{12} - \left(\frac{r_0}{r}\right)^{6}\right] \tag{42}$$

where the parameters ϵ and r_0 have the meaning as indicated in Figure 3.

In correspondence with Marshall and Pople [9], Stephen [10] and Buckingham and Stephen [11], the authors write

$$(\sigma_w)_{pair} = -B\,\overline{F^2} \tag{43}$$

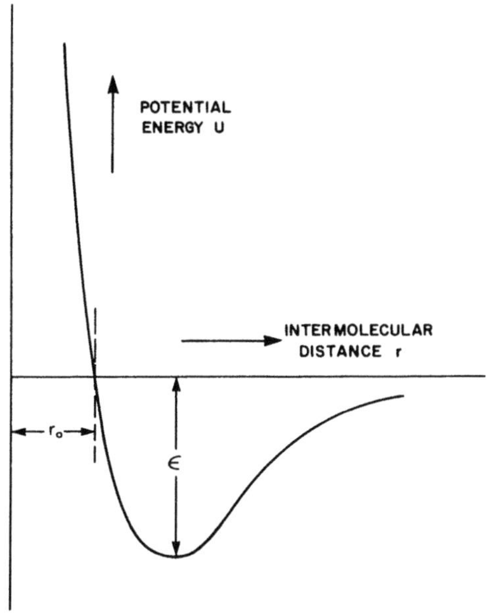

Fig. 3. Lennard-Jones (6–12) potential energy u as function of intermolecular distance r [Eq. (42)] with definition of the parameters ϵ and r_0

where $\overline{F^2}$ relates to the fluctuating field experienced by solute (1) due to the other molecule (2) of the pair. They also showed [15], that approximately one has

$$\overline{F^2} = \frac{3\,\alpha_2 I_2}{r^6} \tag{44}$$

The combination of Eqs. (40) through (44) then leads to:

$$\sigma_w = \frac{(\sigma_1)_w}{V_M} = \frac{-\pi\,NB\,\alpha_2\,I_2}{V_M r_0^3 y^4}\,\mathscr{H}_6(y) \tag{45}$$

where y is defined as

$$y = 2\sqrt{\epsilon/kT}. \tag{46}$$

The functions $\mathscr{H}_n(y)$ have been tabulated by Buckingham and Pople [41]. For mixtures the following combination rules were used;

$$\epsilon = \sqrt{\epsilon_1\,\epsilon_2} \tag{47}$$

$$r_0 = (r_{01} + r_{02})/2 \tag{48}$$

Provided σ_w is indeed found to vary linearly with the density and provided all other molecular parameters are known, Eq. (45) can be used to find the value of B. Figure 4 reproduces such a plot for C_2H_6. In the paper quoted [15], the authors use experimental data on pure CH_4 and pure C_2H_6, combined with similar data on C_2H_4 of Gordon and Dailey [23] to arrive at $B = (1.0 \pm 0.3)\cdot 10^{-18}$ esu, which compared rather well with the value of $0.74\cdot 10^{-18}$ esu calculated for atomic hydrogen [9]. The experimental uncertainty was such that no real distinction could be ascertained between the B values derived for the three individual molecules. It was, therefore, implied that the B parameter found should be the same, (within the error range given) for CH bonds of all hydrocarbons.

In the above theory the parameter B is a pure solute bond parameter, which should be independent of the solvent gas employed. This important aspect was checked by Petrakis and Bernstein [42]. Fluorine-19 resonances were used for this study because it could be expected that the B parameters for X-F bonds would be considerably larger (and therefore more accurately measured) than those for X–H bonds. The results were quite satisfactory. For instance, for CF_4 in six solvents $B\cdot 10^{18} = 16.4 \pm 2.1$ was found (although this excludes the measurements on pure CF_4 which resulted in a 40% lower B value!). Similarly, for SiF_4 in four solvents $B\cdot 10^{18} = 43.5 \pm 5.1$ and for SF_6 in four solvents $B\cdot 10^{18} = 29.5 \pm 2.4$ was found. The standard deviations in these parameters are considerably larger (in absolute terms) than the ± 0.3 quoted for C-H. The experimental accuracy was about the same in either the ^{1}H or the ^{19}F study, but the influence of the molecular parameters such as α_2, I_2, r_0 and ϵ/k is roughly a proportional one. Uncertainties in these can easily add up to 10% in the calculations so that it could be concluded

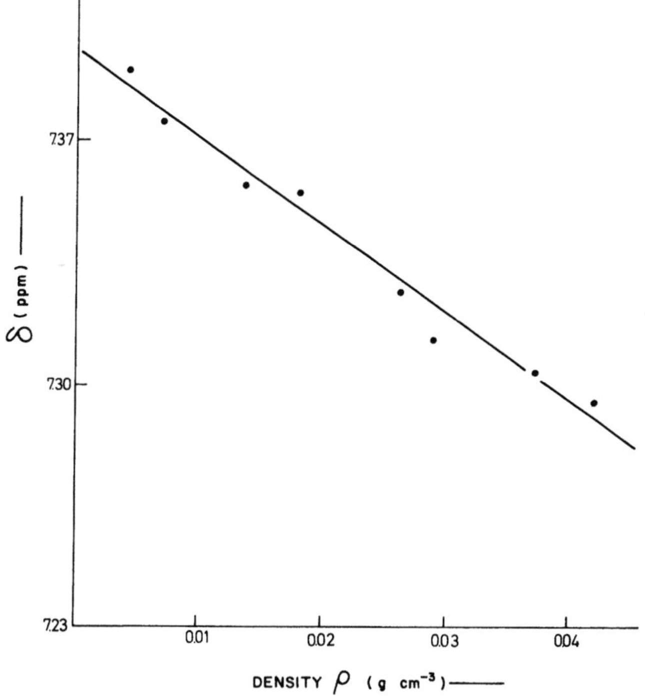

Fig. 4. Chemical shift of ethane gas (relative to liquid benzene as external reference) as function of gas density. After Raynes, Buckingham and Bernstein [*15*]

that the B parameters found were transferable from one solvent to the next. However, such striking exceptions as noted for pure CF_4 (and for SiF_4 in SF_6), cannot be explained on the basis of uncertain other parameters since the same parameters, used in other solute solvent systems, gave results that were at least internally consistent.

There are a good many aspects of the binary collision model for σ_w, such as the constancy of B, the presence or absence of a repulsive term or factor in σ_w, the effect of higher order collisions, the site factor, the temperature dependence of σ_w and the matter of choosing potential parameters that have been studied mainly in connection with the binary gas model. Most of these matters have general implications however. Therefore, rather than under the general heading of the binary gas model, they will be discussed in the respective sections devoted to these subjects. Experiences with this model, applied to *gas-to-liquid* shifts, will be discussed in Chapter 10.

3.2. The Cage Model

Bernstein and Raynes [*43*] have proposed that the total solvent effect can be obtained by taking the interaction of a *pair* and by multiplying this with a suitable coordination number Z, where Z is the number of nearest neighbouring solvent molecules. It was proposed to obtain Z by dividing the surface area of the sphere going through the centres of the solvent molecule in the first coordination layer (i.e. $4\,\pi(r_1 + r_2)^2$) by the effective cross section $(2\,r_2)^2$ of the solvent molecules. This results in

$$Z = \frac{\pi(r_1 + r_2)^2}{r_2^2} \tag{49}$$

The parameters r_1 and r_2 are the effective radii of solute and solvent respectively. An estimate of these can be obtained by assuming that all liquids are close-packed. For such packing one has

$$r^3 = 0.293\ V_M \tag{50}$$

if r is expressed in Å and V_M in cm^3. The use of a close-packed model is at least consistent with the definition of Z as per Eq. (49). For a pure liquid ($r_1 = r_2$), Z becomes equal to $4\,\pi$, which is close to the theoretical $Z = 12$ for close-packing. Also, for $r_1 \ll r_2$ Eq. (49) gives $Z = \pi$ which is close to the theoretical $Z = 4$ for a void in a close-packed system. Combination of Eqs. (43, 44), and (49) results in

$$\sigma_W = \frac{-3\,\pi B\alpha_2 I_2}{r^6} \cdot \frac{(r_1 + r_2)^2}{r_2^2} = \frac{-3\,\pi B\alpha_2 I_2}{(r_1 + r_2)^4 r_2^2} \tag{51}$$

A somewhat simpler form may be obtained for those instances where r_1 and r_2 are not too different. One has then as a first order approximation

$$(r_1 + r_2) \approx 2\,\sqrt{r_1 r_2} \tag{52}$$

which results in

$$\sigma_W = \frac{-3\,\pi B\alpha_2 I_2}{16\ r_1^2 r_2^4} \tag{53}$$

To our knowledge only Rummens, Raynes and Bernstein [17] have tried this cage model, in a discussion of gas-to-liquid σ_W data. Some further details on this model are given in Sections 10.2 and 11.3.

3.3. Kromhout and Linder's Quantum Mechanical Model

In 1969 Kromhout and Linder [44] published a novel and comprehensive theory of σ_W. This model has two unique features. Firstly, they developed a pair interaction term σ_W(pair) on the basis of *two interacting atoms*, rather than on the basis of an atom in an electric field. Secondly, they proposed to use radial distribution functions for the statistical averaging over the entire fluid.

The first part amounts to a direct quantum mechanical calculation of the B parameter. While we will give more details on this in Section 9.5, the major results are reproduced below. For a pair of interacting atoms, Kromhout and Linder find

$$\sigma_w(\text{pair}) = \sigma_0 W\left(\frac{3\,U_1 + 2\,U_2}{U_1(U_1 + U_2)}\right)C, \tag{54}$$

where σ_0 is the shielding of the unperturbed probe nucleus and U_1 and U_2 are the effective excitation energies of solute and solvent atom. The parameter C is given by

$$C = \left[1 - \frac{\underset{j}{\Sigma} <r_j>}{\underset{j}{\Sigma} <r_j^{-1}> \underset{j}{\Sigma} <r_j^2>} + \frac{2\,U_1}{(U_1 + U_2)(3\,U_1 + 2\,U_2)\underset{j}{\Sigma} <r_j^2>}\right] \tag{55}$$

The parameter C is equal to 0.633 for two interaction H atoms, 0.85 for two interacting He atoms, 0.95 for a He atom interacting with any atom that has a 10ev ionisation potential and in general very close to unity for all other pairs. W is the Van der Waals dispersion energy which, in the London formulation, is equal to

$$W = \frac{-3\,\alpha_1\alpha_2 U_1 U_2}{2\,r^6(U_1 + U_2)} \tag{56}$$

Combining Eqs. (54) and (56) and using ionisation potentials rather than effective excitation energies gives

$$\sigma_w(\text{pair}) = \frac{-3\,\alpha_1\alpha_2 C\sigma_0 I_2(3\,I_1 + 2\,I_2)}{2\,r^6(I_1 + I_2)^2} \tag{57}$$

Kromhout and Linder point out that for two H atoms the above leads to $\sigma_w(\text{pair})r^6/\sigma_0 = 24.2$ as compared to 23.86 in Marshall and Pople's calculation on two interacting H atoms [45].

It should be noted that the above does not challenge the basic equation $\sigma_w(\text{pair}) = -B\overline{F^2}$, but it does challenge the use of a B value computed for an atom in a static field in an equation essentially dealing with rapidly *fluctuating* fields. Of course, an approach as described above does eliminate the need for *explicit* use of $\overline{F^2}$.

In order to find $\sigma_w = N\sigma_w(\text{pair})$, Kromhout and Linder write the average Van der Waals interaction energy $\overline{W}$ in terms of a radial distribution function, which leads to

$$\sigma_w = \frac{-6\,\pi N\alpha_1\alpha_2 C\sigma_0 I_2(3\,I_1 + 2\,I_2)}{V_M(I_1 + I_2)^2} \int_0^\infty \frac{\rho(Z)dZ}{r^4} \tag{58}$$

which for pure liquids reduces to

$$\sigma_W = \frac{-15 \pi N \alpha^2 \sigma_0 C}{2 V_M} \int_0^\infty \frac{\rho(Z)dZ}{r^4} \tag{59}$$

Using the previous finding [46] that the integral in Eqs. (58) and (59) was very close to $2/3\, r_0^3$ for liquid Xenon and extrapolating this to hold for all liquids they find for the general case

$$\sigma_W = \frac{-4 \pi N \alpha_1 \alpha_2 \sigma_0 C I_2 (3 I_1 + 2 I_2)}{V_M r_0^3 (I_1 + I_2)^2} \tag{60}$$

or, for pure liquids,

$$\sigma_W = \frac{-5 \pi N \alpha^2 \sigma_0 C}{V_M r_0^3} \tag{61}$$

For atomic liquids the above theory (both parts of it) may well be very attractive. The predicted gas-to-liquid shifts for He, Ne, Kr and Xe come out to 0.042, 2.4, 144 and 365 ppm respectively which values are very reasonable. It should be pointed out, however, that the above theory does not include any paramagnetic contribution to the medium shift. The applicability to *molecules* is therefore uncertain. Another drawback is the proposed simplification of the radial distribution function integral; the factor $2/3\, r_0^3$ is a constant for a given system and therefore part of the temperature dependence of σ_W is lost. On the other hand, some simplification is necessary because the integral in question is usually very untractable (both theoretically and experimentally), particularly for molecular liquids. For non-spherical molecules the radial distribution is not even entirely relevant since orientation will also enter then.

Nevertheless Kromhout and Linder calculate a gas-to-liquid shift of 0.22 ppm for CH_4 which compares well with the experimental value of 0.20 obtained by Gordon and Dailey [23]. For CF_4 they calculated 5.5 ppm, which the authors compare with a value of 2.0 ppm obtained by extrapolating available gas phase data to the liquid density. However, as has been observed frequently [17], if not universally, the actual gas-to-liquid shift is considerably larger (almost 2 x) than such extrapolated data (see section 10.2).

It might be concluded perhaps, that Kromhout and Linder's theory gives the correct order of magnitude and that this theory is fully worthwhile of further exploration. Evaluation of the radial distribution function integral, incorporation of an appropriate site factor (see Chapter 6) and inclusion of paramagnetic terms seem almost unsurmountable obstacles at the present. The most worthwhile achievement is likely the theory for σ_W(pair) which amounts to a better definition of either the effective F^2 or of B. These aspects will be discussed in more detail in Chapters 5 and 9 respectively.

3.4. A Preliminary Comparison of the Various Models Suggested for σ_w

Using a few simple approximations, all models discussed above can be rewritten as product functions of polarisabilities, ionisation potentials and molar volumes V_1 and V_2 of solute and solvent. These approximations are

(i)   the use of the Lorentz-Lorenz equation [Eq. (27)] to eliminate $n^2 - 1$;
$$n_2^2 - 1 = 4\,\pi(n_2^2 + 2)\alpha_2 N/3\,V_2$$

(ii)   the substitution of $n^2 = 2$ in terms like $(n^2 + 2)$ and $(2\,n^2 + 1)$.

(iii)   the Onsager relation [Eq. (23)] to eliminate the cavity radius
$a_i^3 = 0.238\,V_i/N$.

(iv)   the London relation [Eq. (20)] to eliminate resonance frequencies $h\nu = I$.

(v)   the relation $I_1 + I_2 \approx 2\,\sqrt{I_1 I_2}$.

(vi)   the use of $r^3 = 0.955\,V/N$ [Eq. (50)] for effective cage radii, and
$r_0^3 = 0.955\,N/V$ (from corresponding-states theory $b_0 \simeq 2\,V = 2\,\pi N r_0^3/3$)
for Lennard Jones diameters.

(vii) replacement of algebraic by geometric averages [Eq. (52)] for radii r and
diameters r_0.

The resulting simplified formula are collected in Table 7.

Table 7. Functional comparison of the proposed expressions for σ_w

	Polarisability (cm³)	Ionisation potential (erg)	Molar volumes (cm³)	Proportionality coefficient
Continuum				
Linder, Eq. (19)	α_2	$I_1^{1/2} I_2^{1/2}$	$V_1^{-1} V_1^{-1}$	$-10.5\,BN^2$
De Montgolfier Eq. (29)	$\alpha_2^4 \alpha_1^{-1}$	I_1	V_2^{-4}	$-1180\,kBN^4$
Rummens, Eq. (35)	$\alpha_2^2 \alpha_1$	I_1	$V_1^{-2} V_2^{-2}$	$-1180\,kBN^4$
Binary Collision *Gas:*				
Raynes, Buckingham & Bernstein, Eq. (45)	α_2	I_2	$V_1^{-1/2} V_2^{-3/2}$	$-0.411\,BN^2 \dfrac{\mathscr{H}_6(y)}{y^4}$
Cage:				
Bernstein & Raynes, Eq. (53)	α_2	I_2	$V_1^{-3/2} V_2^{-4/3}$	$-18.8\,BN^2$

It is seen, therefore, that the various models have a great deal of similarity, but that they are, amongst others, fundamentally different in their dependence on the molar volumes of solute and solvent. However, one should note that for most molecules the ratio α_i/V_i is close to a universal constant (approximately 0.5). Taking the product of the polarisability and the molar volume columns of Table 7 and assuming a constant ratio α_i/V_i the molar volume dependences of the five models listed in Table 7 become V_1^{-1}, $V_1^1(!)$, V_1^{-1}, $V_1^{-1/2} V_2^{-1/2}$ and $V_1^{-2/3} V_2^{-1/3}$ respectively. Again it is found that De Montgolfier's model leads to an unacceptable result.

The difference in dependence on ionisation potential will be discussed in Chapter 5.

The binary gas model distinguishes itself by explicitly taking into account an intermolecular potential (Figure 3), which leads to a temperature dependent coefficient $\mathscr{H}_6(y)/y^4$. By contrast the continuum and cage models are implicitly assuming potentials as given in Figure 5.

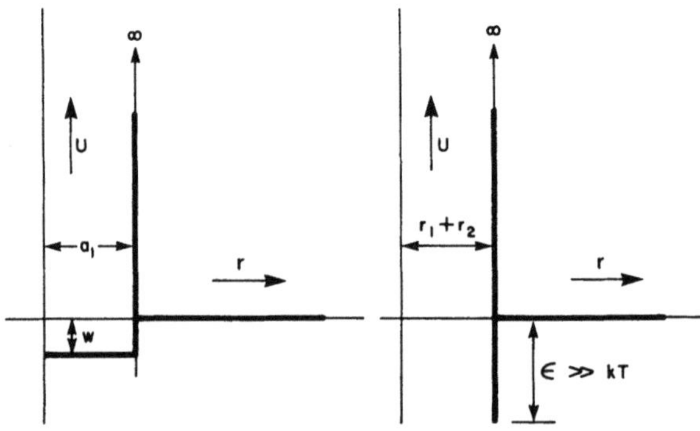

Fig. 5. At the left the potential in a continuum theory; the cavity with radius a has an energy $W = -\alpha_1 E^2/2$ relative to all parts of the continuum; at $r = a_1$ the potential is infinitely large. At the right the potential corresponding to the cage model; only solvent molecules at the first coordination layer, at distance $(r_1 + r_2)$ are counted; these molecules must therefore have a much lower potential than all others at $r > (r_1 + r_2)$. For $r < (r_1 + r_2)$ the potential becomes infinitely large

Chapter 4. Other Experimental Proton σ_w Data

In the previous chapters illustrative experimental data were given wherever it seemed appropriate, but much more data have in fact been published. These have all been collected in the present chapter, to provide a basis for discussion of various points of detail regarding σ_w in the chapters that follow. Unless noted to the contrary, experimental data have been recalculated using better susceptibility data, if such adjustment seemed warranted. In a few cases calculated σ_w were also available; these have been included.

At the end of the chapter a list of gas phase chemical shifts is given. Strictly speaking these data have no bearing on σ_w at all. However, such gas phase data are indispensable when externally (and even internally) referenced experimental data are to be converted to gas-to-liquid data (see Chapter 15).

Table 8. $-\sigma_w$(ppm) at 25 °C according to Lumbroso, Wu and Dailey [27]. Data not corrected for finite pressure of the gases or for finite concentration in the solution

Solute \ Solvent	C_6H_{12}	Dioxane	CCl_4
CH_4	0.28	0.36	0.40
C_2H_6	0.21	0.23	0.31
C_2H_4	0.24	0.34	0.37
cyclo-propane	0.27	0.32	0.38
$C(CH_3)_4$	0.23	–	0.29
TMS	0.27	–	0.36

Table 9. $-\sigma_w$(ppm) at 30 °C for $X(CH_3)_4$ compounds according to Rummens [17]. Data obtained by extrapolation to zero pressure gas and infinite dilution

Solute \ Solvent	$C(CH_3)_4$	$Si(CH_3)_4$	$Ge(CH_3)_4$	$Sn(CH_3)_4$	$Pb(CH_3)_4$
$C(CH_3)_4$	0.217	0.185	0.215	0.222	0.277
$Si(CH_3)_4$	0.255	0.228	0.262	0.270	0.322
$Ge(CH_3)_4$	0.260	0.228	0.260	0.275	0.325
$Sn(CH_3)_4$	0.280	0.250	0.280	0.297	0.350
$Pb(CH_3)_4$	0.285	0.258	0.287	0.302	0.358

Table 10. $-\sigma_w$(ppm) at 30 °C obtained by Raynes [17]. Data are extrapolated to zero pressure and infinite dilution

Solute \ Solvent	$C(CH_3)_4$	$Si(CH_3)_4$	CCl_4	$SiCl_4$	C_6H_{12}	$n–C_7H_{16}$
$C(CH_3)_4$	0.222	0.212	0.307	0.307	0.220	–
$Si(CH_3)_4$	0.240	0.205	0.322	0.240	0.233	0.215
C_6H_{12}	0.187	0.182	0.267	0.185	0.202	0.170
$n–C_7H_{16}(C\underset{\sim}{H}_2)$	0.192	0.148	0.225	0.152	0.152	0.145
C_6H_6	0.268	0.243	0.397	0.272	0.203	0.200

Other Experimental Proton σ_w Data

Table 11. $-\sigma_w$(ppm) at 37 °C gas-to-solution of some simple gases according to Dayan and Widenlocher [47]. Gas pressure 2–5 atm, both in the gas phase and in the solution; corrected for susceptibility of solvent only

Solute	Solvent		
	CCl_4	$n-C_6H_{14}$	C_6H_{12}
H_2	0.20	0.01	0.00
CH_4	0.08	−0.01	−0.06
C_2H_6	0.02	–	−0.09
C_2H_4	0.06	−0.01	−0.06
C_2H_2	0.31	0.08	0.05

Note: Table 11 shows some *upfield* Van der Waals shifts. We feel that this must be due to faulty experimentation. Comparison of the above data for CH_4, C_2H_6, and C_2H_4 in CCl_4 and C_6H_{12} with those of Table 8 shows that the latter σ_w data are consistently larger by about 0.3 ppm.

The error is likely due to an erroneous σ_b correction for TMS; Dayan [48] shows liquid TMS (corrected for σ_b) upfield from gaseous TMS by 0.057 ppm, while in fact it is 0.228 ppm downfield (see Table 9). The difference is 0.285 ppm, which fits with the above empirical observation.

Table 12. $-\sigma_1$ and $-(\sigma_1)_w$ (in $cm^3 mole^{-1}$ ppm) for CH_4, C_2H_6, C_2H_4 and H_2 at approximately 25 °C. Except for the date of Oldenziel (whose own experimental χ_v data were used), all susceptibility corrections were recalculated using the most recent susceptibility data [19]. Therefore, for $CH_4(\sigma_1)_b = -36.4$, for $C_2H_6(\sigma_1)_b = -56.1$, for $C_2H_4(\sigma_1)_b = -39.4$ and for H_2 and HD$(\sigma_1)_b = -8.35$ (ppm cm^3 $mole^{-1}$) was used

	ref.	$-(\sigma_1)$	$-(\sigma_1)_w$
	a	44	7.6 ± 2
	b	42	5.6 ± 2
	c	43	6.6 ± 3
CH_4	d	46.9	7.7 ± 1
	f	45.8	9.4
	g	41.8	5.4 ± 2
	b	74	17.9 ± 2
C_2H_6	c	65	8.9 ± 4
	e	73.3	17.2 ± 1
	a	50	10.6 ± 2
C_2H_4	c	48	8.6 ± 3
	d	51.5	10.2 ± 1
H_2, HD	c	13	4.6 ± 2

(a) Gordon and Dailey [23]
(b) Raynes, Buckingham and Bernstein [15]
(c) Dayan and Widenlocher [49]
(d) Oldenziel [50]
(e) Rummens [24]
(f) Meinzer [51]
(g) Mohanty and Bernstein [52]

Table 13. $-\sigma_W$(ppm) of TMS and C_6H_{12} according to Jouve [53]. Temperature not stated, but probably 35 °C. (Varian A-60). Original data were obtained with external standard. These were converted to $-\sigma_W$ data, using the gas-to-liquid shifts $\sigma_W = -0.233$ for TMS in C_6H_{12} and $\sigma_W = -0.267$ for C_2H_{12} in CCl_4 (Table 10) and using the χ_V data given by Jouve

Solute	Solvent				
	n-hexane	C_6H_{12}	CCl_4	decalin	$CCl_2{=}CCl_2$
TMS	0.230	(0.233)	0.267	0.270	0.315
C_6H_{12}	0.244	–	(0.267)	–	0.294

Table 14. $-\sigma_W$(ppm) at 20 °C of TMS according to Friedrich [54]. Original data were given relative to external reference of TMS (5%) in CCl_4. Data have been recalculated to gas-to-liquid basis using $\sigma_W = -0.233$ ppm for TMS in C_6H_{12} (see Table 10), and using the χ_V data as given by Friedrich

Solvent	$-\sigma_W$
C_6H_{12}	(0.233)
decalin	0.250
n-C_6H_{14}	0.230
n-C_7H_{16}	0.247
n-C_8H_{18}	0.276
i-C_8H_{18}	0.248
p-dioxane	0.284
$CCl_2{=}CCl_2$	0.247
CCl_4	0.299

Table 15. $-\sigma_W$(ppm) at 35 °C according to Raza [55], Raza and Raynes [56] and Raynes and Raza [59]

Solute	Solvent						
	C_6H_{12}	C_5H_{10}	$SiEt_4$	$SnEt_4$	$SnMe_4$	$SiCl_4$	CCl_4
C_6H_{12}	0.143	0.165	0.145	–	0.175	0.187	0.265
C_5H_{10}	0.157	0.163	0.165	0.178	0.185	0.203	0.295
$Si(CH_2CH_3)_4$	0.140	0.143	0.153	–	0.153	0.162	0.257
$Sn(CH_2CH_3)_4$	–	0.170	–	0.207	0.187	0.205	0.282
$C(CH_3)_4$	0.187*)	0.195*)	0.197	0.205	0.205	0.223	0.345
$Si(CH_3)_4$	0.282	0.312	0.250	0.260	0.267	0.298	0.375
$Sn(CH_3)_4$	0.293	0.295	0.292	0.310	0.310	0.325	0.433
CH_4	–	–	0.305	0.317	0.322	0.347	0.472
n-$C_6H_{14}(CH_3)$	0.197	0.217	–	0.208	0.213	0.220	0.297
$(CH_3)_2C{=}C(CH_3)_2$	0.217	0.215	0.218	0.232	0.230	0.237	0.340

Table 15. (continued)

Solute \ Solvent	C_6H_{12}	C_5H_{10}	$SiEt_4$	$SnEt_4$	$SnMe_4$	$SiCl_4$	CCl_4
p-$C_6H_4(CH_3)_2$	–	–	0.242	0.267	0.268	0.283	0.423
$1,3,5$-$C_6H_3(CH_3)_3$	–	–	0.250	0.272	0.278	0.295	0.417
$CH_3C{\equiv}CCH_3$	0.270	0.273	0.277	0.300	0.288	0.318	0.477
C_6H_6	0.257	0.260	0.240	0.273	0.277	0.293	0.443
p-$C_6H_4F_2$	–	–	0.253	0.287	0.290	0.315	0.492
p-$C_6H_4(CH_3)_2$	–	–	0.200	0.230	0.227	0.245	0.340
$1,3,5$-$C_6H_3(CH_3)_3$	–	–	0.183	0.202	0.205	0.220	0.292

*) Values from [55]; in the subsequent publication [56] values are quoted which are 0.017 ppm smaller.

Table 16. $-\sigma_w$(ppm) of large sphere-like solutes in sphere-like solvents at 38 °C, according to Louman [57] and Rummens and Louman [58]. Data are extrapolated to zero-pressure gas and infinite dilution. Calculated data according to Eq. (76) with $K_6^g = 1.65$ [17]

Solute \ Solvent			CCl_4		C_6H_{12}		$SiBr_4$		TMS	
			exp	calc	exp	calc	exp	calc	exp	calc
$C(CH_2CH_3)_4$	CH_3	1	0.220	0.217	0.143	0.166	0.217	0.210	0.123	0.113
	CH_2	2	0.153	0.152	0.098	0.117	0.142	0.152	0.080	0.082
$Si(CH_2CH_3)_4$	CH_3	3	0.242	0.257	0.165	0.196	0.238	0.248	0.122	0.131
	CH_2	4	0.190	0.179	0.127	0.138	0.185	0.179	0.088	0.095
$Sn(CH_2CH_3)_4$	CH_3	5	0.265	0.335	0.185	0.254	0.252	0.325	0.147	0.166
	CH_2	6	0.222	0.226	0.162	0.174	0.212	0.228	0.120	0.117

Gas-to-pure-liquid σ_w shifts

	CEt_4		$SiEt_4$		$SnEt_4$	
	exp	calc	exp	calc	exp	calc
(CH_3)	0.184	0.093	0.145	0.105	0.184	0.138
(CH_2)	0.139	0.072	0.109	0.081	0.156	0.104

Table 17. $-\sigma_w$(ppm) at 38 °C of non-spherical solutes according to Louman [57] and Rummens and Louman [58]. Data are extrapolated to zero-pressure gas and infinite dilution. Calculated data according to Eq. (76) with $K_6^g = 1.65$ [17]

Solute \ Solvent			CCl_4		C_6H_{12}		$SiBr_4$		TMS	
			exp	calc	exp	calc	exp	calc	exp	calc
p-xylene	CH_3	9	0.397	0.466	0.260	0.349	0.378	0.424	0.208	0.225
	CH	10	0.300	0.242	0.197	0.185	0.298	0.237	0.165	0.125
mesitylene	CH_3	11	0.365	0.400	0.238	0.300	0.352	0.370	0.175	0.196
	CH	12	0.252	0.217	0.177	0.166	0.245	0.214	0.122	0.113

Table 17. (continued)

Solute	Solvent		CCl$_4$ exp	CCl$_4$ calc	C$_6$H$_{12}$ exp	C$_6$H$_{12}$ calc	SiBr$_4$ exp	SiBr$_4$ calc	TMS exp	TMS calc
butyne-2	CH$_3$	20	0.460	0.497	0.275	0.373	0.389	0.454	0.228	0.242
trans-butene-2	CH$_3$	13	0.372	0.407	0.237	0.308	0.359	0.375	0.203	0.202
	CH	14	0.255	0.247	0.148	0.190	0.268	0.240	0.135	0.130
2,3-pentadiene	CH$_3$	18	0.392	0.485	0.263	0.363	0.379	0.440	0.210	0.235
	CH	19	0.328	0.285	0.223	0.217	0.322	0.275	0.197	0.197
2,4-dimethyl-2,3-pentadiene	CH$_3$	15	0.332	–	0.228	–	0.472	–	0.177	–
hexyne-3	CH$_3$	16	0.347	0.845	0.218	0.625	0.343	0.738	0.168	0.382
	CH$_2$	17	0.338	0.455	0.225	0.343	0.347	0.429	0.168	0.222
iso-butane	CH$_3$	7	0.298	0.315	–	–	0.265	0.301	0.167	0.161
	CH	8	0.302	0.245	–	–	0.272	0.241	0.175	0.129

Table 18. Proton gas phase chemical shifts $-\sigma_0^*$ (ppm) relative to ethane

Compound	$-\sigma_0^*$	foot note
Si(CH$_3$)$_4$	−0.900	a
	−0.882	b
	−0.875	c
Sn(CH$_3$)$_4$	−0.860	b
	−0.835	c
CH$_4$	−0.767	d
	−0.75	e
Ge(CH$_3$)$_4$	−0.748	c
Si(C$\underset{\sim}{H}_2$CH$_3$)$_4$	−0.242	c
	−0.223	f
Pb(CH$_3$)$_4$	−0.182	c
C(CH$_2$C$\underset{\sim}{H}_3$)$_4$	−0.035	f
C$_2$H$_6$	0.000	–
Sn(C$\underset{\sim}{H}_2$CH$_3$)$_4$	0.032	f
(C$\underset{\sim}{H}_3$)$_2$CH$_2$	0.05	a
(C$\underset{\sim}{H}_3$CH$_2$)$_2$	0.05	a
(C$\underset{\sim}{H}_3$)$_2$CH	0.050	f
(C$\underset{\sim}{H}_3$CH$_2$CH$_2$)$_2$	0.055	g
C(C$\underset{\sim}{H}_3$)$_4$	0.092	b
	0.095	c
Si(CH$_2$C$\underset{\sim}{H}_3$)$_4$	0.150	c
	0.152	f
	0.165	g
(=CCH$_2$C$\underset{\sim}{H}_3$)$_2$	0.215	f

Table 18. (continued)

Compound	$-\sigma_0^*$	foot note
$(CH_3)_2CH_2$	0.43	a
$Sn(CH_2CH_3)_4$	0.373	f
	0.373	g
$(CH_3CH_2CH_2)_2$	0.487	g
n-heptane CH_2	0.517	c
C_2H_2	0.60	e
	0.42	a
C_6H_{12}	0.63	a
	0.635	c
	0.650	b
$(=C=CHCH_3)_2$	0.693	f
C_5H_{10}	0.702	b
$CH_3C\equiv CCH_3$	0.702	b
	0.718	f
$trans$-$(CH_3CH=)_2$	0.708	f
$(=C=C(CH_3)_2)_2$	0.742	f
$(=C(CH_3)_2)_2$	0.763	b
$CH(CH_3)_3$	0.875	f
$(\equiv CCH_2CH_3)_2$	1.218	f
$1,2,4,5$-$C_6H_2(CH_3)_4$	1.270	g
$1,3,5$-$C_6H_3(CH_3)_3$	1.308	g
	1.325	f
$1,4$-$C_6H_4(CH_3)_2$	1.338	g
	1.342	f
SiH_4	2.25	e
H_2	3.35	h
	3.55	d
	3.58	i
HD	3.51	d
$(=C=CHCH_3)_2$	4.057	f
C_2H_4	4.38	d
	4.43	e
$trans$-$(=CHCH_3)_2$	4.570	f
$1,3,5$-$C_6H_3(CH_3)_3$	5.823	g
	5.863	f
$1,2,4,5$-$C_6H_2(CH_3)_3$	5.943	g
$1,4$-$C_6H_4F_2$	5.957	g
$1,4$-$C_6H_4(CH_3)_2$	6.063	g
	6.095	g
C_6H_6	6.308	g
	6.33	a
	6.345	c

a) At ambient temperature (35 °C?). Measured at partial pressure below 1 atm relative to internal TMS at equally low partial pressure. Conversions based on the internal shift of CH_4-TMS of 0.133 ppm and $CH_4-C_2H_6$ internal shift of 0.767 ppm (see d). Data from Dayan [48] and Dayan and Widenlocher [49].

b) At 35 °C obtained by extrapolating from plots of chemical shift *versus* temperature. Original measurements were with 1 atm of partial pressure for the solute against internal C_2H_6, the latter at about 2 atm. Published data, which were given relative to CH_4 on the basis of a $CH_4-C_2H_6$ shift of 0.75 ppm, have been reconverted to the original C_2H_6 basis, using the same conversion. Data of Raynes and Raza [59].

c) At about 80 to 120 °C measured at 1 to 2 atm of pressure relative to external liquid acetone, plus a separate determination of liquid acetone versus 10 atm C_2H_6. All gases were corrected for bulk susceptibility; the ethane was also corrected for σ_w, so that the reference is to zero pressure ethane. Data from Rummens, Raynes and Bernstein [17]. Note that in this reference the sign for σ_0 of CH_3 of $Si(CH_2CH_3)_4$ is wrong, but that the difference $\sigma_0(CH_2)-\sigma_0(CH_3)$ is correct.

d) At ambient temperature (35 °C?). Data obtained by extrapolation to zero pressure, with liquid TMS as intermediate reference. Data from Dayan [48] and Dayan and Widenlocher [49].

e) At 22 °C, measured internally at approximately 15 atm of total pressure, at 7.5 atm partial pressure each. Data from Schneider, Bernstein and Pople [12].

f) At 180 °C. Measured with internal reference to C_2H_6 (3–5 atm partial pressure). Partial pressure of solute variable from 1 to 6 atm. Data from Louman [57] and Rummens and Louman [58]. See also Table 30 for the temperature dependencies of some of the gas phase shifts.

g) At 175 °C. Measured by internal reference to C_2H_6 with both gases at "relatively low pressures". Data of Raza and Raynes [56] given relative to CH_4, which have been readjusted back to the C_2H_6 basis by adding −0.750 ppm.

h) Ambient temperature (20 °C (HR 60) or 35 °C (A 60)?). Measurements by reference to liquid TMS by substitution. Pressure of gases unknown. Only CH_4 and H_2 were measured. (4.12 ppm difference). Correction of 0.77 has been applied for the $CH_4-C_2H_6$ difference. Data of Chaigneau, Dayan and Widenlocher [60].

i) At 34 °C. Measured with a mixture of 9 atm of CH_4 and 1 atm of H_2, which gave 4.35 ± 0.15 ppm between these two signals. No corrections for intermolecular effects made. Listed value obtained by subtracting 0.77 ppm to account for $CH_4-C_2H_6$ shift. Data from Hinderman and Cornwell [61].

Chapter 5. The Physical Nature of the Field $\overline{F^2}$ and of the Associated Excitation Energy

In the foregoing several methods were described to evaluate the time-averaged square of the electric field as experienced by the solute. It is perhaps useful to stress that not only the *methods* of evaluating $\overline{F^2}$ are different, but that in fact fields of quite different nature are evaluated with various methods.

5.1. The Bothner-By Method

In this method the classical expression for the energy of a polarisable particle (the solute) in an electric field is equated to the semi quantum mechanical interaction energy as *per* the London theory. In the latter theory the Schrödinger equation is solved for a system of two identical one-electron harmonic oscillators. The interaction energy

$$\phi_L = -\frac{3}{4}\frac{h\nu\alpha^2}{r^6} \tag{62}$$

indicates how much lower the energy of the system of two coupled oscillators *in its ground state* is as compared to the sum of the ground state energies of the two oscillators when separated. The interaction energy is negative because of the built-in assumption that the two oscillating dipoles are head-to-tail. This in turn is equivalent to saying that the instantaneous moment of the one molecule *induces* an oscillation in the other. The resulting dispersion force is then said to be an induced-dipole-induced-dipole force. The equivalent role of solute and solvent molecule is even more apparent in the formulation for non-identical oscillators (the Drude model) in which

$$\phi_L = -\frac{3}{2}\frac{\alpha_1\alpha_2}{r^6}\frac{\nu_1\nu_2}{\nu_1+\nu_2} \tag{63}$$

To equate this to a classical expression $\phi_c = \alpha_1\overline{F^2}/2$ appears of doubtful accuracy, if only because the London-Drude model requires a zero-point energy, which has no equivalent in the classical energy. However, if one accepts this equality, then it should be clear that the $\overline{F^2}$ so derived [(see Eq. (8)] is the field at the solute, due to the moment at the solvent molecule. The oscillation of the latter is, however, modified due to the dipolar coupling. The modified moment and the modified frequency of the solvent molecule are always expressable in terms of the moments, frequencies and polarisabilities of the two separate molecules. This is the reason (and the *only* reason) why solute parameters like ν_1 and α_1 enter into the equation for F^2. The field so derived does not include any direct contribution from the electrons of the solute itself, nor does it include a reaction field contribution. Note that the field is that due to *one* solvent molecule and that it is calculated at the centre of mass of the solute if r stands for the inter-centre-of-mass distance. Summation of this F^2 over all solvent molecules faces fundamental problems as will be detailed in the next section. Finally, ν_2 is a ground state oscillation frequency of the solvent. A good approximation would be to substitute $\nu_2 = (E_1-E_0)_2/h$ where (E_1-E_0) is the main transition energy. The use of $\nu_2 = I_2/h$ [Eq. (20)] would lead to an overestimate of at least a factor of 2. Note also that the consistent use of the London-Drude model requires the use of $2\,\nu_1\nu_2/(\nu_1 + \nu_2)$ rather than ν if solute and solvent are different. Bothner-By's calculation was for a pure liquid, however.

5.2. The Raynes-Buckingham-Bernstein (RBB) Method

In this method the detour via dispersion energy and the mixture of classical and quantum mechanical theory is altogether avoided. These authors write directly:

$$\overline{F^2} = \frac{2}{r^6} \langle 0 | m_2^2 | 0 \rangle \tag{64}$$

for the average squared fluctuating field a distance r from an unperturbed solvent molecule in the ground state.

Second order perturbation result for the polarisability is then given as

$$\alpha_{2xx} = 2 \sum_{n \neq 0} \frac{\langle 0 | m_{2x} | n \rangle \langle n | m_{2x} | 0 \rangle}{(E_n - E_0)_2}$$

$$= 2 \langle 0 | m_{2x} | \sum_{n \neq 0} \frac{| n \rangle \langle n |}{(E_n - E_0)_2} m_{2x} | 0 \rangle \tag{65}$$

For a complete set of orthonormal wave functions one has $\sum_n | n \rangle \langle n | = 1$ which under the assumption that each $(E_n - E_0)$ term may be replaced by an *average* excitation energy $\overline{(\Delta E)}_2$ leads to

$$\alpha_{2xx} = -\frac{2}{\overline{(\Delta E)}_2} \langle 0 | m_{2x}^2 | 0 \rangle \tag{66}$$

or, assuming electrically isotropic molecules,

$$\alpha_2 = \frac{2}{3 \, \overline{(\Delta E)}_2} \langle 0 | m_2^2 | 0 \rangle \tag{67}$$

Combination of Eqs. (64) and (67) then results in

$$\overline{F^2} = \frac{3 \, \alpha_2}{r^6} \sum \frac{(E_n - E_0)_2}{| n \rangle \langle n |} \approx \frac{3 \, \alpha_2}{r^6} \overline{(\Delta E)}_2 \approx \frac{3 \, \alpha_2 I_2}{r^6} \tag{68}$$

where $\overline{(\Delta E)}_2$ is a better approximation than the substitution $h\nu_2 = I_2$, while in turn the use of I_2 for $\overline{(\Delta E)}_2$ is a better approximation than the substitution $h\nu_2 = I_2$ of the London-Drude model. Under all circumstances $\overline{(\Delta E)}_2$ will be intermediate between $(E_1 - E_0)_2$ and I_2 and probably fairly close to I_2 for complex molecules with a great number of energy levels close to the ionisation limit, all having finite contributions to $\overline{F^2}$.

Three critical comments can be made *re* the RBB method. Firstly, the appearance of a factor 2 in Eq. (65) is due to the use of a rather old-fashioned and now disused perturbation theory based upon the following

$$\mathcal{H} = \mathcal{H}^{(0)} + \frac{\lambda}{1!} \mathcal{H}^{(1)} + \frac{\lambda^2}{2!} \mathcal{H}^{(2)} + \ldots \tag{69}$$

as compared to the presently more commonly used

$$\mathscr{H} = \mathscr{H}^{(0)} + \lambda\,\mathscr{H}^{(1)} + \lambda^2\,\mathscr{H}^{(2)} \tag{70}$$

It is the factorial 2! in the second order perturbation term of Eq. (69) which gave rise to the factor 2 in the expression for α of Eq. (65). Yet if cannot be stated that Eq. (65) is wrong; provided that it is used consistently the same results should be obtained. However, it is inconsistent to use Eq. (67) together with Eq. (64). While in Eq. (70) λ^2 can be identified with $\overline{F^2}$, in Eq. (69) and in its consequences it is $\overline{F^2}/2$ that is evaluated. Alternately, if Eq. (64) is maintained, then the factor 2 in Eqs. (65–68) should be withdrawn. Either way it appears that $\overline{F^2}$ of Eq. (68) is twice too large and that it should be replaced by

$$\overline{F^2} = \frac{3\,\alpha_2(\overline{\Delta E})_2}{2\,r^6} \approx \frac{3\,\alpha_2 I_2}{2\,r^6} \tag{71}$$

As may be seen, the result appears now the same as with the Bothner-By method [Eq. (8)], but with the significant difference that the $\overline{F^2}$ of the RBB method [Eq. (71)] is supposed to hold also for mixtures, whereas Eq. (8) is only valid for pure liquids, to be modified according to Eq. (63) for mixtures.

Our second criticism is based on a corollary of the above, in that the RBB method is based on the field of an *isolated* solvent molecule, whereas the required field is that at a solute molecule at close distance; the lack of any perturbation on the solvent due to the solute creates almost certainly a substantial error in this method. The difference is essentially that between I_2 and $2\,I_1 I_2/(I_1 + I_2)$. Since $2\,I_1 I_2/(I_1 + I_2)$ can be either smaller or larger than I_2 this results in a further mis-estimate of uncertain sign in $\overline{F^2}$ if Eq. (68) is used. This point will be elaborated in section 5.6.

The third objection is one that becomes relevant as soon as pair interaction expressions such as Eq. (71) are used to calculate a σ_w effect due to *all* the surrounding solvent molecules. This therefore applies not only to the binary gas model (section 3.1), but also to the cage model (section 3.2) as well as to any summation method that might be proposed using Bothner-By's model (section 5.1). In all of these one takes a solvent molecule and calculates the electric field due to that oscillating solvent molecule at the solute molecule; one squares this field and averages this over the oscillation of that solute-solvent pair, and *then* the $\overline{F_i^2}$'s of all pairs are added in a scalar fashion. Our question is, however, "How does the solute molecule know that it should add the fields only after they have been squared and averaged?". The answer seems to be that this is wrong; the $\overline{F^2}$ at the solute is the time-averaged square of the total field $\overline{F} = \Sigma\,\overline{F_i}$ due to all solvent molecules. But the individual fields $\overline{F_i}$ are not additive in a scalar fashion; both because of their vector character, as well as because of the different phases of the oscillations of each solvent molecule, there will be a large degree of cancellation. Even without having worked out the statistics it would appear that $(\overline{F_{\text{total}}})^2$ cannot be much larger than $\overline{F_i^2}$ for a single pair. However, now it follows that the shielding σ_w can hardly be due to such direct fields; only if the individual fields $\overline{F_i}$ would be parallel and exactly in phase could a substantially enough $\overline{F^2}$ be expected. The reaction field theory (see following two sections) seems to fulfil these conditions.

5.3. Linder's Model

The derivation of Linder's $\overline{F^2}$ has already been given [Eqs. (10–18)]. It appears to be an extension of the Drude model, in that it does include a term $\nu_1 \nu_2/(\nu_1 + \nu_2)$, but it is actually different in two major aspects. Firstly the $\overline{F^2}$ derived is taken at the solute and due to the entire surrounding, the latter being considered continuous. While the usage of a continuum may seem questionable, it has the undisputable advantage of not requiring an integration over pair interactions (not to mention the worries about higher order collisions) such as is inherent in any pair model. The second distinctive difference is the phrasing in terms of a reaction field. The solute molecule with its spontaneous oscillating moment, when brought into the cavity, polarises the continuum, which in turn creates a reaction field at the solute. After the initial unstable period the oscillating moment and the reaction field reach their equilibrium (averaged) values. These final values of $\overline{m^2}$ and $\overline{R}$ are then used in expressions for the dispersion energy. The nature of the $\overline{F^2}$ so derived is therefore that of a pure reaction field, without any contribution from the direct fields, quite opposite to the $\overline{F^2}$ models discussed in the previous two sections. In Linder's model all solvent molecules, or rather all volume elements of the continuum, are in resonance with the chosen solute molecule; the induced moments in all volume elements and their reaction fields are parallel and exactly in phase. Actually, it is sufficient that this condition exist in a sphere of dimensions larger than the chosen cavity, but smaller than the bulk system. Additionally, such resonant conditions are assumed to be very short-lived, after which similar situations arise around other centers. The attractiveness of the reaction field model lies in the above mentioned phase- and direction coherence. This is corroborated by the large cohesion energy of non-polar liquids; the cohesion energy of polar liquids is usually only slightly larger than expected on the basis of dispersion forces only. The polar contributions are commonly small and become only comparable to the dispersive contributions if both solute and solvent have large permanent moments (≥ 2.0 D).

One might ask why Linder, having arrived at an expression for the reaction field R^* [Eq. (12)], did not simply square this to obtain a usable $\overline{F^2}$. The reason appears to be that the averaging over the distributions $\rho(\nu_i)$ and $\rho(\nu_j)$ becomes quite untractable. It is only in the expression for the work W [Eq. (14)] that a neat trick ($\alpha_i g_j \simeq \alpha_j g_i$) makes this averaging feasible.

In Linder's theory the frequencies that are supposed to occur are spontaneous oscillations which initially are supposed to have a certain range. No indication is given of the width of this frequency band; ultimately an averaging process is carried out and the averaged frequencies are identified with the ground state oscillations of isolated harmonic oscillators. This process of averaging does lead to a factor $h\nu_1 \nu_2/(\nu_1 + \nu_2)$ which also occurs in the London-Drude theory and is there due to the mutual perturbation of oscillators at close distance. At least formally therefore, Linder's theory does appear to incorporate this aspect.

Finally it might be repeated that by definition the reaction field is constant over the cavity; a site factor is therefore inconsistent with this model. However, as duly admitted by Linder, the reaction field theory can only be precise for a cavity considerably larger than molecular dimensions. For a cavity of molecular dimensions it is no longer tenable to consider the cavity wall as if polarised in a macroscopic fashion. A second condition for a constant reaction field is the isotropy of the solute particle; this point has been challenged by De Montgolfier (see next section).

5.4. De Montgolfier's Method

De Montgolfier's field is essentially the reaction field due to the spontaneously oscillating *solute* [see Eqs (33) and (30)]. Therefore, it is closely related to the field as described by Linder. The starting points are identical and most of the differences are simplifications to frame the models in terms of easily available parameters. This has been discussed already in section 2.2. The only basic difference is that De Montgolfier considers the electrical anisotropy of chemical bonds; as a result of this the reaction field is not constant over the cavity. A discussion of this particular kind of site effect will be given in section 6.3.

5.5. The Method of Kromhout and Linder

All previously discussed methods attempt to find an expression for $\overline{F^2}$, which then is multiplied by $-B$ to obtain σ_w (pair).

Kromhout and Linder's theory is different because it directly calculates σ_w (pair) resulting from the interaction of *two* atoms. The calculation of Marshal and Pople [45] on two interacting H atoms is along the same line.

The $\overline{F^2}$ field does not enter at all in Kromhout and Linder's theory and therefore one cannot say much about its nature. One could of course equate their expression for σ_w (pair) [Eq. (54)] to $-B\overline{F^2}$ and so obtain an expression for an equivalent $\overline{F^2}$. However, this expression contains B, a parameter which again is foreign to this theory. All that one can extract is an expression for B, if $\overline{F^2}$ is evaluated by other means. We will further discuss this matter in section 9.3.

5.6. An Empirical Test

All methods discussed above indicate a dependence on oscillator frequencies for $\overline{F^2}$, usually in the form of either ν_2 or of $\nu_1\nu_2/(\nu_1 + \nu_2)$ (or in terms of I_2 or $I_1I_2/(I_1 + I_2)$). If ν_1 and ν_2 (or I_1 and I_2) are not too different then the following approximation is acceptable:

$$\frac{2\nu_1\nu_2}{\nu_1+\nu_2} \approx \sqrt{\nu_1\nu_2} \quad \text{or} \quad \frac{2I_1I_2}{I_1+I_2} \approx \sqrt{I_1I_2} \tag{72}$$

Mohanty and Bernstein [52] in their ^{19}F study on CF_4, SiF_4 and SF_6 concluded on empirical statistical grounds that using $\sqrt{I_1I_2}$ rather than I_2 gave superior correspondence between calculated and experimental Van der Waals shifts. No further details were given, however.

A demonstration [62a] of a similar effect can be given, however, using the matrix of 25 σ_w's of $X(CH_3)_4$ compounds studied by Rummens, Raynes and Bernstein [17]. In Table 19 the experimental results are reproduced followed by the calculated values using I_2 (upper calculated number), followed by calculations based on $\sqrt{I_1 I_2}$ (lower calculated number).

Table 19. Comparison of the use of I_2 versus $\sqrt{I_1 I_2}$ in the expression for σ_w, tested out for $X(CH_3)_4$ systems. The σ_w data (in ppm) were calculated using Eq. (76) including a scale factor, determined by a least squares fit.

Solute \ Solvent	CMe$_4$ exp	CMe$_4$ calc	SiMe$_4$ exp	SiMe$_4$ calc	GeMe$_4$ exp	GeMe$_4$ calc	SnMe$_4$ exp	SnMe$_4$ calc	PbMe$_4$ exp	PbMe$_4$ calc
CMe$_4$	0.217	0.247	0.185	0.200	0.215	0.217	0.222	0.227	0.277	0.245
		0.247		0.207		0.230		0.252		0.275
SiMe$_4$	0.255	0.233	0.228	0.217	0.262	0.232	0.270	0.247	0.322	0.265
		0.227		0.217		0.240		0.267		0.288
GeMe$_4$	0.260	0.260	0.228	0.240	0.260	0.260	0.275	0.273	0.325	0.297
		0.245		0.232		0.260		0.287		0.312
SnMe$_4$	0.280	0.290	0.250	0.272	0.280	0.290	0.297	0.313	0.350	0.330
		0.262		0.250		0.278		0.313		0.333
PbMe$_4$	0.285	0.302	0.258	0.278	0.287	0.303	0.302	0.320	0.358	0.357
		0.268		0.255		0.288		0.317		0.357

The improvement using $\sqrt{I_1 I_2}$ is noticeable; the original matrix shows deviations ranging from -0.057 to $+0.030$ ppm, whereas with the $\sqrt{I_1 I_2}$ model this reduces to -0.033 to $+0.030$ ppm. The standard deviation decreases from ± 0.021 to ± 0.017 ppm. More significantly, the original calculated matrix shows rather systematically negative deviations (average -0.015 ppm) in the upper right corner elements and positive deviations (average $+0.010$ ppm) in the lower left corner. This systematic bias disappears upon using $\sqrt{I_1 I_2}$.

Chapter 6. The Site Factor

6.1. The Solute Site Factor of Rummens *et al.*

Most molecular interaction theories and models put all molecular properties at the centre of mass of the molecule, essentially treating the molecule as a point except for the interaction potential which endows a size on the molecules. The intermolecular distance is invariably defined as the distance between the centres of mass. In NMR, where measurements are done on nuclei (^{1}H, ^{19}F) which are usually on the periphery of the molecules such assumption seems not too well warranted; a fact recognized, but still neglected in the early work of Bothner-By [7] and of others after him [12, 18, 35, 55].

Rummens and Bernstein have put this idea into a quantitative form [16]. It was reasoned that it is the solute-nucleus-to-solvent-centre distance R rather than the inter-centre distance r which is important and that one should therefore use (see also Figure 6)

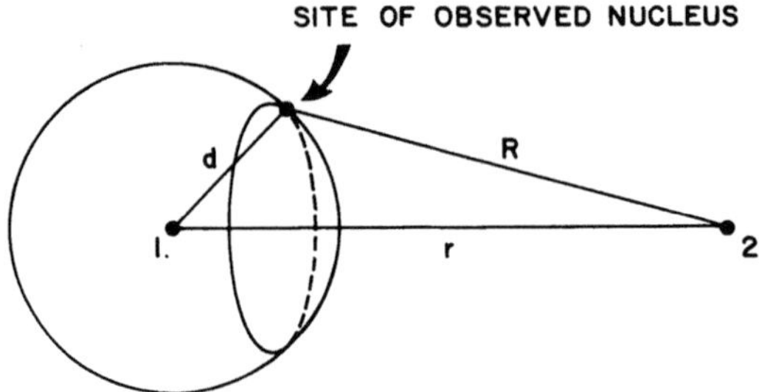

Fig. 6. The site of the observed nucleus in the solute molecule 1 in respect to the center of the solvent molecule 2. Note that whereas $< R > = r$, in general $< R^{-n}> \neq r^{-n}$

$$\langle \overline{F^2} \rangle = 3\, \alpha_2 I_2\, \langle R^{-6} \rangle \tag{73}$$

where the averaging is over all the positions of the measured nucleus relative to the solvent molecule. Assuming absence of any rotational restriction for the solute gives the following relation between $\langle R^{-6} \rangle$ and r^{-6}

$$\langle R^{-6} \rangle = r^{-6}[(1 + q^2)/(1-q^2)^4]$$ (74)

where q is defined as

$$q = \frac{d}{r}$$ (75)

d being the distance from measured nucleus to the centre of mass of the solute molecule. While the above is for a *pair* of interacting molecules, insertion into the binary collision gas theory results in

$$\sigma_w = \frac{-\pi B N \alpha_2 I_2}{V_M r_0^3 y^4} \mathscr{H}_6(y) \left\{ 1 + 5 q_0^2 \frac{\mathscr{H}_8(y)}{\mathscr{H}_6(y)} + 14 q_0^4 \frac{\mathscr{H}_{10}(y)}{\mathscr{H}_6(y)} \right.$$

$$\left. + 30 q_0^6 \frac{\mathscr{H}_{12}(y)}{\mathscr{H}_6(y)} + 55 q_0^8 \frac{\mathscr{H}_{14}(y)}{\mathscr{H}_6(y)} + \ldots \ldots \right\}$$ (76)

Since the ratios $\mathscr{H}_n(y)/\mathscr{H}_6(y)$ are almost constant in the y range employed, the site factor [i.e. the expression between brackets in Eq. (76)] can be written as

$$S_6^g = 1 + 3.45 q_0^2 + 7.42 q_0^4 + 12.9 q_0^6 + 19 q_0^8 + \ldots \ldots \ldots$$ (77)

where q_0 is defined as

$$q_0 = \frac{d}{r_0}$$ (78)

r_0 being the appropriate Lennard-Jones (6–12) parameter. The site factor S_6^g is a rapidly increasing function of q_0, being unity for $q_0 = 0$ and about 3 for $q_0 = 0.5$ (see also Figure 13, section 16.3).

Putting this model to work on earlier experimental data by Bernstein and coworkers on the σ_w effects of CH_4, C_2H_6, C_2H_4, SiF_4, CF_4 and SiF_6, Rummens and Bernstein then recalculated the B parameters. It turned out that not only were these new B parameters somewhat smaller, more importantly they proved much better transferable between solute-solvent combinations. The standard deviations for the B parameters all decreased by 40 to 50%.

A better test yet was provided by Rummens, Raynes, and Bernstein in a study of gas-to-liquid proton shifts [17]. Using initially a matrix of 25 gas-to-liquid shifts (see Table 19) of binary systems made up from $X(CH_3)_4$ (X=C, Si, Ge, Sn, and Pb) molecules it was shown that the experimental results could be reproduced with a standard deviation of ±0.022 ppm (in addition a scale factor $K_6^g = 1.65$ was used). This is the more remarkable since any other model which does not incorporate a site factor, predicts a virtually constant gas-to-liquid shift for these 25 systems ($\alpha_2 I_2/V_1 V_2$ happens to be virtually constant in this series) while in fact they range over a full factor of 2. The experimental and calculated results have been given already in Table 19; a graphical demonstration is given in Figure 7.

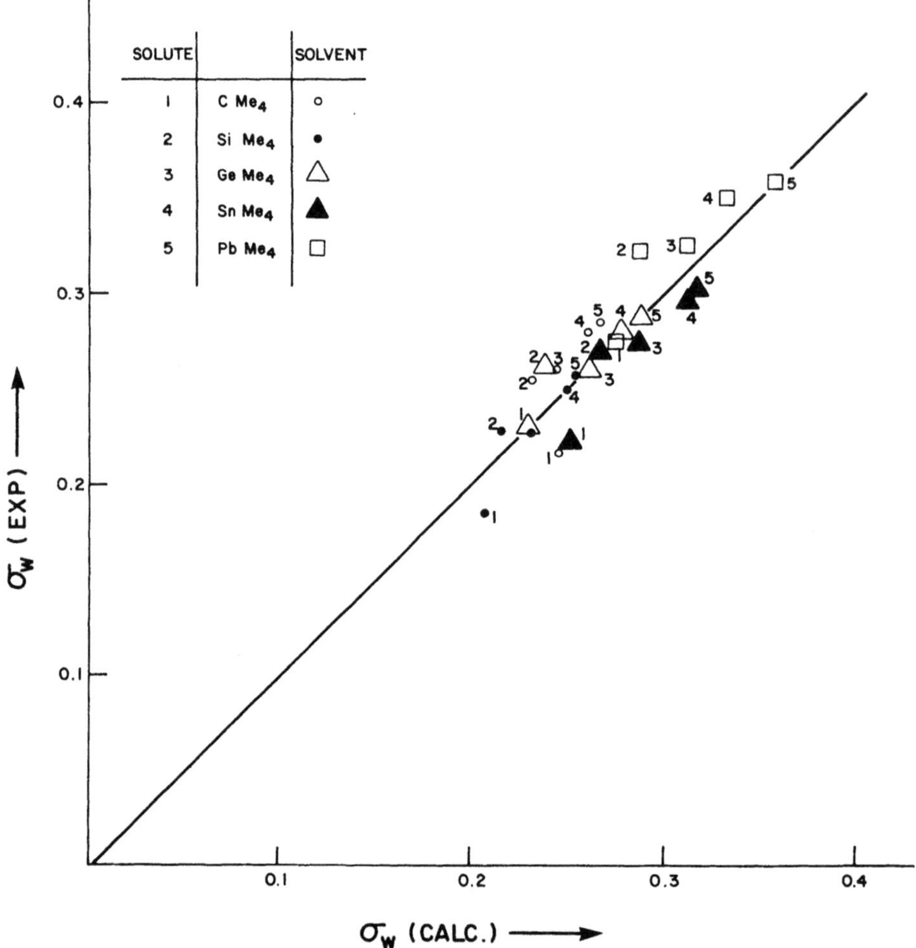

Fig. 7. Test of the site factor theory. Calculate σ_W data according to the binary collision gas model Eqs. (76) and (77), with scale factor $K_6^g = 1.65$ and with the additional refinement of replacing I_2 by $\sqrt{I_1 I_2}$ (see section 5.6). Experimental data and molecular parameters as used by Rummens, Raynes and Bernstein [17]

In the same paper [17] the site factor (with the same scale factor) was tried out on 52 other gas-to-liquid shifts of the non-polar/non-polar isotropic type. Measurements of Bothner-By [7] (Table 1), Buckingham, Schaefer and Schneider [18] (Table 2) and of Lumbroso, Wu and Dailey [27] (Table 18) were used while some new experimental data as given in Table 10 were added. In spite of the wide variety of compounds, many of which were only spherical in a rough approximation and in spite of large (up to 0.05 ppm) discrepancies between measurements of the same shifts reported by different authors, a standard deviation of ±0.04 ppm was obtained using the binary gas model and the site factor as per Eq. (77).

An even more direct proof of the existence of the site factor was also given. Gas-to-liquid shifts of $Si(CH_2CH_3)_4$ in six isotropic non-polar solvents were given. the data showed that the medium shifts for CH_3 were always larger than the corresponding CH_2 shifts by 0.03 to 0.05 ppm as they should indeed, following the site factor model (Note: the actual reported gas-to-liquid shifts are apparently considerably in error as was found later by Rummens and Louman [57, 58] upon repeating these experiments. The error does not influence the CH_2/CH_3 shift *differences,* however (see also footnote c in Table 18).

Similar effects were found by Raynes and Raza [59] for $Si(OCH_2CH_3)_4$. Again the gas-to-liquid shifts for CH_3 were always found to be larger than those of CH_2. However, we must point out that molecules like $Si(OCH_2CH_3)_4$ have a strong permanent electric dipole moment, a fact overlooked by Raynes and Raza. Their data for nonpolar/nonpolar isotropic systems were given in Table 15. Raza [55] and Raynes and Raza [59] do remark that many trends reflect the "degree and exposure" of the measured protons to the solvent, with larger exposure resulting in larger σ_w's. Unfortunately, they do not give any quantitative considerations.

The site factor, as developed in Eqs. (73–78) is not without limitations. It assumes that the solute molecule is effectively spherical with the measured nuclei on spherical shells around the centre of mass. Even for the pseudo-spherical molecules such as $Si(CH_2CH_3)_4$ this is not true; in any of its possible conformations at least one of the CH_2 groups will be just as exposed as the CH_3 groups. The alternative interpretation, which assumes that the rotation of the solute molecule as a whole is much faster than the translational motions of the solvent molecules, is equally unattractive for large molecules. This kind of limitation is demonstrated by the experimental σ_w data on $X(CH_2CH_3)_4$, with X = C, Si and Sn as given in Table 16. Indeed, $\sigma_w(CH_3)$ is always larger than $\sigma_w(CH_2)$ for each molecule; also $\sigma_w(CH_3)$ and $\sigma_w(CH_2)$ increase both in the direction X = C $\rightarrow$ Sn, as predicted, but the difference $\sigma_w(CH_3)-\sigma_w(CH_2)$ *decreases* in that same direction, contrary to the expectations based on Eq. (77) for the site factor. Apparently with increasing size of the central atom the solute molecule can more easily be penetrated by the solvent molecules and hence the ethyl fragments will increasingly act independently. This can rather strikingly be demonstrated in the following way. If one plots the $\sigma_w(CH_3)$ and the $\sigma_w(CH_2)$ shifts for the $X(CH_2CH_3)_4$ solutes in any solvent versus some site factor indicator, such as the parameter q_0, one obtains two straight lines which converge; the chemical shift corresponding to the intersection is equal to the σ_w of ethane in that same solvent [57].

If the solute molecule is far from spherical the site factor model as given can also be expected to fail; the solvent environment around one end of a stick-like solute molecule should become more and more independent of what happens at the other end as the solute molecule is made longer and longer. This is demonstrated by the data of Table 17 with disc-like molecules such as mesitylene and rod-like solutes such as hexyne-3. For "inner" protons the calculated σ_w values are still reasonable, but for "outer" protons σ_w values are calculated that are invariably too large.

Rummens and Louman [58] have attempted to account for the above mentioned two problems by introducing an empirical site factor S_e

$$S_e = 1 + \frac{3.74\, q_0^2 (1 + q_0^2)}{(1 + 28\, q_0^8)} \tag{79}$$

They claim that, whereas Eq. (78) gives only reliable results up to $q_0 = 0.4$, Eq. (79) would extend the validity range to $q_0 = 0.55$. While Eq. (79) may have its usefulness we should point out that it forms only a surrogate solution; the problem of tying nuclear site to size and shape of solute molecules can never be treated adequately using a single parameter q_0.

Finally there is a mathematical objection to the development of Eq. (76). As De Meyer [62a] has found, this series does not converge. Subsequent terms may get smaller for a certain number of terms and approach closely to a zero value, but later terms will be larger again and the series actually diverges. The reason for this is the use of the Lennard-Jones potential which has finite values down to $r = 0$; therefore there is a small but finite probability that the centre of mass of the solvent molecule will coincide with the nuclear site. To avoid this problem a Kihara-type hard core potential has to be used where the potential goes to infinity at a distance $l = l_1 + l_2$ (l_1 and l_2 being the radii of the hard cores of the solute and solvent respectively). For spheres the potential can be given by

$$u = 4\, \epsilon^* \left[\left(\frac{\rho_0}{\rho} \right)^{12} - \left(\frac{\rho_0}{\rho} \right)^{6} \right] \tag{80}$$

with $\rho = r - (l_1 + l_2)$.

In the integration over the r coordinate, the integral now is from $l \to \infty$ rather than from $0 \to \infty$; the integrals $\mathcal{H}_n(y)$, which could formally be written as a sum over gamma functions now must be described in terms of incomplete gamma functions. There are several criteria as to the choice of the core radii l_i, but the most important one (to obtain convergence) is to choose $l_1 + l_2 > d$. Fortunately, it could be shown that Eq. (77), if truncated after five terms as indicated, gives normally results within 2% of the correct values.

A new approach for σ_w based on a potential involving general ellipsoidal hard cores is presently being developed [62b]. This results not only in a much improved definition of the site effect, but brings automatically into account the anisotropy of molecular geometry, of the polarisability and of the susceptibility. Only a few initial results are available at present; these results are in the correct direction, but no definite assessment is presently possible.

6.2. The Solute-Solvent Site Factor of Raynes

Raynes in a paper on interacting CH_4 molecules [63] has pointed out that not only is there a *solute* site factor, but in principle also a similar site factor for the solvent. In his theory the contribution to the dispersion field of the five atoms of a CH_4 solvent molecule are considered separately. Using the binary gas model formalism, Raynes finds (for interacting CH_4 molecules) a site factor

$$S_{6,6}^{g} = 1 + 0.803 \frac{\mathscr{H}_{8}(y)}{\mathscr{H}_{6}(y)} + 0.832 \frac{\mathscr{H}_{10}(y)}{\mathscr{H}_{6}(y)} + 0.445 \frac{\mathscr{H}_{12}(y)}{\mathscr{H}_{6}(y)} + \ldots \tag{81}$$

The site factor S_{6}^{g} [from Eq. (76)] as applied to CH_4 in CH_4 would give the following;

$$S_{6}^{g} = 1 + 0.413 \frac{\mathscr{H}_{8}(y)}{\mathscr{H}_{6}(y)} + 0.095 \frac{\mathscr{H}_{10}(y)}{\mathscr{H}_{6}(y)} + 0.017 \frac{\mathscr{H}_{12}(y)}{\mathscr{H}_{6}(y)} \tag{82}$$

It is seen that $S_{6,6}^{g}$ is considerably larger than S_{6}^{g}. This would in turn lead to much higher calculated σ_w values. However, there are two compensating factors. Firstly, even if this technique were adopted, the semi-empirical B parameters would have to be recalculated, using the same model, which would compensate for most of the effect. Secondly, the field of an assembly of dipoles (at the five atoms of solvent CH_4) outside a sphere enveloping these dipoles can be written as the sum of the fields of an ideal dipole, ideal quadrupole, etcetera. This solvent site factor is therefore equivalent to introducing electric quadrupole (and higher order poles) effects leading to terms in r^{-8}, r^{-10}, etcetera in the dispersion energy. To be consistent such terms should therefore be added to the potential energy expression and possibly also in the σ_w (pair) interaction term. As will be shown in Chapter 8, the effect of higher order dispersion terms is already largely incorporated in the B parameter and in the potential parameters ϵ and r_0. Finally, the derivation of Raynes contains a similar error as indicated in the previous section; the series of Eq. (81) does not converge. This time the magnitude of the deviation is much worse because of the dominating influence of the very short H(solute) H(solvent) distances. Furthermore, if one would now introduce a hard core potential to improve the calculation, the entire solvent site problem vanishes because such a hard core is effectively equivalent to atomic contributions smeared out over the surface or over the volume of the core, as has been shown by Pitzer [64].

6.3. The Solute Site Factor of De Montgolfier

An entirely different solute site factor idea has been used by De Montgolfier. In his first paper dealing with this subject [31] he argues that the reaction field at a specific atom of the solute is not equal to the Onsager reaction field but proportional to it as follows [see also Eq. (33)]

$$\overline{F_{H}^{2}} = k \, \overline{R^{2}} \tag{83}$$

in which k is larger than unity. The argument is that no molecule is a true point-molecule; therefore an atom of a molecule in the Onsager cavity would not only experience the homogeneous Onsager field, but *in addition* the fields of the bond dipoles of the solute molecules induced by that Onsager reaction field. Partly because of the geometry of a molecule and partly because of the anisotropy of bond polarisabilities this extra effect does not add up to zero. De Montgolfier then calculates $k \simeq 2.5$ for CH_4 in good agreement [31] with the empirical scale factor he required to obtain correspondence between the experimental CH_4 shifts in a number of solvents and the calculated σ_W shifts according to his own continuum theory (see section 2.2).

Although the principle of a molecule being not a point and of one part of the solute experiencing some reaction field contribution due to another part of the same solute is unquestionably correct there are some major criticisms to be made. Firstly, we do not agree with leaving out of consideration the bond in which the atom under consideration resides. De Montgolfier states that this is done because "Buckingham's theory takes these electrons already into account" [31]. However, this should refer only to the proper value of B to be used. Inclusion of this bond would of course mean the addition of an overwhelmingly large extra term upsetting the entire calculation.

Secondly, as De Montgolfier himself shows abundantly [31], the calculation of k depends very sensitively on the choice of location of the dipoles in each bond. Essentially one ends up guessing this proper distance and therefore guessing the value of k with an error margin perhaps as large as 100%.

More fundamentally we object to first calculating the homogeneous Onsager reaction field and then *adding* heterogeneous effects. Surely, by definition the reaction fields is always a total effect which even in Onsager's theory includes the effects of a polarisable solute. It would be more proper to start out the reaction field calculations by placing an assembly of polarisable bonds in a cavity and try to obtain the total reaction field from there. Such a method would automatically include the effects of the solute bonds, but would not have the normal Onsager reaction field as a common basis.

Although in one paper [30] De Montgolfier finds empirically $k \simeq 2.5$ for CH_4 and $k \simeq 10$ for cyclopentane (whence the large difference?) he argues in a subsequent paper that these factors must be closely the same [33]. The reasoning is interesting because it involves redefining the cavity as just containing the bond under consideration; the rest of the solute has become part of the solvent continuum. For any bond type, say C–H, the expression $k \cdot B \Delta E/\alpha$ [see Eq. (29)] then simply refers to parameters of that bond (and not of the entire solute). However, since B, ΔE and α must be very similar for any CH bond, he concludes that k must be a constant for any C–H bond, contrary to his findings. One may remark here that placing all the rest of the solute molecule in the solvent continuum is in contradiction to the rest of his theory including the calculations of k!

In his further work, De Montgolfier does not use any calculated k parameters. In comparing calculated and experimental shifts (σ_W and σ_E) he ends up extracting empirical values for the product kB. With no prior knowledge of the values of B the coupling with another equally unknown parameter seems rather superfluous. Even if B were known to some accuracy, k would still be different for each solute and each bond type and would have to be determined empirically as a fitting parameter.

In summary therefore it seems that there is no experimental evidence for the site effect as advanced by De Montgolfier. That is not to say that such a factor does not exist; it may well exist but its presence cannot be established. This situation is very similar to that of the repulsion effect (Chapter 7) and the effect of higher order dispersion terms (Chapter 8) and also bears on the basic uncertainty regarding the value of B parameters (Chapter 9).

Chapter 7. The Repulsion Effect

7.1. The Repulsion Formulation in the Binary Gas Collision Model; Application to Effects on Protons

At very short intermolecular distance the interaction energy becomes repulsive due to overlap. Apart from being of importance in the Boltzmann weighting factor in the statistical mechanical gas model, one may ask whether these overlap forces also have a direct effect on the shielding. Marshall and Pople [45] did calculate the screening parameter σ of two interacting H atoms using Heitler-London wave functions. Below, the overlap contributions σ_{rep}, as calculated by Marshall and Pople are given, converted to ppm on the basis of σ_0 = 17.8 ppm for an isolated H atom.

distance R	σ_{rep}
(a.u.)	(ppm)
2	−1.104
3	0.000
4	+0.077
5	+0.036
6	+0.012
7	+0.004

Compared and combined with the dispersive contribution to the shielding, which were also calculated by Marshall and Pople it followed that the total effect is deshielding at large distances, but that there is an overall shielding effect between 4 and 7 Bohr radii. At shorter distances yet the combined effect is deshielding again. As Rummens and Bernstein showed [16] this function of σ vs r is very similar to that of the *interaction energy vs r* (as calculated by Hirschfelder and Linnett [65]). They therefore postulated that for protons at least, σ would be proportional to the *total* interaction energy. If the energy can be written as

$$\phi = -4\epsilon \left[\left(\frac{r_0}{r} \right)^6 - \left(\frac{r_0}{r} \right)^{12} \right] = -4\epsilon \left(\frac{r_0}{r} \right)^6 \left[1 - \left(\frac{r_0}{r} \right)^6 \right] \tag{84}$$

then, by analogy, one would have

$$\sigma_w \, (\text{pair}) = -B \, \overline{F^2} \left[1 - \left(\frac{r_0}{r} \right)^6 \right] \tag{85}$$

which, after insertion into Eq. (40) leads to

$$\sigma_w = \frac{-\pi N B \alpha_2 I_2}{V_M r_0^3 y^4} \, \mathscr{H}_6(y) \left[1 - \frac{\mathscr{H}_{12}(y)}{\mathscr{H}_6(y)} \right] \tag{86}$$

Since $\mathscr{H}_{12}(y)/\mathscr{H}_6(y)$ is nearly a constant (about 0.43) the correction term will be a constant too. This leads to an ambiguous situation since the parameter B is not known to a sufficient degree of exactness. If Eq. (86) is used rather than Eq. (45) in the semi-empirical B determination, all that will result is a new B value, larger by a factor $1/0.57$. It appears therefore impossible to determine whether a repulsion effect exists or not.

Rummens and Bernstein also considered the combined effect of repulsion and site factor. They did this by using $<\overline{F^2}>$ as given by Eq. (73) and substitution into Eq. (85). The result then is

$$\sigma_w = \frac{-\pi N B \alpha_2 I_2}{V_M r_0^3 y^4} \, \mathscr{H}_6(y) \left[1 - \frac{\mathscr{H}_{12}(y)}{\mathscr{H}_6(y)} + 5 \, q_0^2 \left\{ \frac{\mathscr{H}_8(y) - \mathscr{H}_{14}(y)}{\mathscr{H}_6(y)} \right\} \right.$$
$$\left. + 14 \, q_0^4 \left\{ \frac{\mathscr{H}_{10}(y) - \mathscr{H}_{16}(y)}{\mathscr{H}_6(y)} \right\} + 30 \, q_0^6 \left\{ \frac{\mathscr{H}_{12}(y) - \mathscr{H}_{18}(y)}{\mathscr{H}_6(y)} \right\} + \ldots \right] \tag{87}$$

Actually one might question whether it would not be more consequential to use

$$\sigma_w \text{ (pair)} = -B \left\langle \overline{F^2} \left[1 - \left(\frac{r_0}{r} \right)^6 \right] \right\rangle = -3 \, B \alpha_2 I_2 \left\langle \left[\frac{1}{r^6} - \frac{r_0^6}{r^{12}} \right] \right\rangle \tag{88}$$

which leads to

$$\sigma_w = \frac{-\pi N B \alpha_2 I_2}{V_M r_0^3 y^4} \left\{ \mathscr{H}_6(y) - \mathscr{H}_{12}(y) \right\} \left[1 + 5 \, q_0^2 \, \frac{\mathscr{H}_8(y)}{\mathscr{H}_6(y)} + \right.$$
$$\left. + 14 \, q_0^4 \, \frac{\mathscr{H}_{10}(y)}{\mathscr{H}_6(y)} + 30 \, q_0^6 \, \frac{\mathscr{H}_{12}(y)}{\mathscr{H}_6(y)} + \ldots \right] \tag{89}$$

All these variations have essentially the same effect; experimental precision and accuracy is insufficient to single out any of the proposed repulsion terms as being more proper than the others. They would all appear equally adequate and the only result would be a series of B parameter sets, from which it would be difficult to chose. Exactly the same kind of duality arises if Eq. (85) is used as a repulsion effect basis in the cage model (section 3–2) because both r_0^3 and r^3 are proportional to the the molar volume.

Rummens, Raynes and Bernstein in their study of gas-to-liquid shifts [17] experienced the above sketched dilemma and "solved" it simply by abandoning the repulsion effect altogether. However, they did find that the temperature dependence of σ_w is more specific in its sensitivity to the repulsion effect (see section 11.2).

54

7.2. Repulsion Effects in ^{19}F σ_w Shifts

For nuclei other than protons the repulsion effect may well be drastically different. This may best be illustrated by the ^{19}F work of Mohanty and Bernstein [52]. They measured CF_4, SiF_4, SF_6 and CH_4 not only as pure gases, but also in the solvent gases CF_4, SiF_4, SF_6, CH_4, Kr and Xe. Furthermore they did this at a number of temperatures, usually over a range of 100 °C for each and every binary combination. They report that when the original RBB theory [see Eq. (45)] was used, two major discrepancies were found. Firstly, the calculated temperature dependence of σ_w was invariably much too low as compared to the experimental data and secondly, the B parameters so derived for each solute differed substantially per solvent (up to a factor of two). Because of the high accuracy of the data and of the large magnitude of B parameters (at least for ^{19}F), the differences in B parameter could not be attributed to experimental uncertainties. The use of a Drude-type theory, using essentially $2 I_1 I_2/(I_1 + I_2)$ rather than I_2, gave only a minor improvement. Inclusion of higher order dispersion terms or any site-repulsion effects [Eqs. (86, 87)] did not give substantial improvement and were therefore subsequently dropped.

Mohanty and Bernstein then tried the following expression for σ_w

$$\sigma_w = \frac{-\pi NBI_1 I_2 \alpha_2}{y^4 r_0^3 V_M (I_1 + I_2)} \, \mathscr{H}_6(y) \left[1 - C \frac{\mathscr{H}_m(y)}{\mathscr{H}_6(y)} \right] \tag{90}$$

This is similar to Eq. (84) except that a repulsion term

$$(\sigma_w \text{ pair})_{rep} = + \frac{c}{r^m} \tag{91}$$

is used, with $C = c(I_1 + I_2)/3 \, B\alpha_2 I_1 I_2 r^{m-6}$, rather than

$$(\sigma_w \text{ pair})_{rep} = + \frac{B\overline{F^2} r_0^6}{r^6} = \frac{3 \, B\alpha_2 I_2 r_0^6}{r^{12}} \tag{92}$$

The parameters C and m were empirically determined so as to fit all available experimental data. It turned out, that by suitable choice of C and m not only the temperature dependence of σ_w could be fitted, but that the universality of B as a solute parameter could be restored. We reproduce Mohanty and Bernstein's results below in Table 20.

Table 20. B, C and m parameters (with standard errors) according to Mohanty and Bernstein [52] following Eq. (90)

Solute	$B \times 10^{18}$ esu	C	m
CF_4	262 ± 20	2.33 ± 0.04	17 ± 1
SiF_4	690 ± 47	2.34 ± 0.03	16 ± 1
SF_6	782 ± 67	2.62 ± 0.03	16 ± 1

We may note that the C and m parameters are almost universal constants, independent of solute, solvent and temperature. This argues strongly against a repulsion effect that modifies $\overline{F^2}$ such as inherent in Eqs. (85) and (92). It rather seems that repulsion has to be treated as an independent phenomenon, as indicated in Eq. (91). It may also be noted that the high value of $m = 16$ to 17 seems quite realistic. It has often been noted that while the Lennard-Jones (6–12) potentials is quite good for atoms, a higher exponent for repulsion appears more appropriate for poly-atomic molecules. Note also that the B parameters are very large; some factor 10 to 20 times larger than those obtained with the simpler RBB theory without repulsion. It would follow thus that the repulsion effect for ^{19}F would be always very large, and in fact always more than 90% of the dispersion effect. Here lies a further possible test; even a rather crude quantum mechanical model should be able to calculate a B parameter for ^{19}F with sufficient accuracy to differentiate between the two different sets. Such calculations have only been done for hydrogen and rare-gas atoms. Jameson, Jameson and Gutowsky [66] calculate $B = 5.6 \cdot 10^{-18}$ esu for Ne which atom is not very different from F. But this B value differs so much from those of Table 20 that the latter now look unrealistically high.

Mohanty and Bernstein also noted that the extracted B parameters for CH_4 [Eq. (45) RBB method] varied notably with the solvent. The variations were large (values ranging from 0.53 to 2.10) and certainly outside experimental error. Apparently because these measurements were done only at room temperature, no attempt was made to find C and m for protons. This is a great pity, because one would very much like to know the relative importance of the repulsion effect with protons.

For further discussion of Bernstein and Mohanty's study see sections 9.6, 10.1 and 11.2.

7.3. Repulsion Effects in ^{129}Xe

In several respects the ^{129}Xe work of Jameson, Jameson and Gutowsky (JJG) [66] is similar to that of Mohanty and Bernstein. ^{129}Xe was measured in a number of solvent gases. When the semi-empiricial RBB method was tried, B values for ^{129}Xe ranging from 318 to $837 \cdot 10^{-18}$ esu were found, again pointing to some deficiency in the theory with discrepancies well beyond the experimental uncertainty levels. The authors did calculate a B value for ^{129}Xe ($337.3 \cdot 10^{-18}$ esu), calculated $(\sigma_1)_w$ values from these for each binary system and noted that these were noticeably smaller (about 40–60%) than the experimental σ_1 data. The remaining part is therefore deshielding, so that one has apparently a downfield repulsion effect. To calculate the repulsion contribution they used a second order perturbation approach with

$$\lambda^2_{rep} = f\, U_{rep}/\Delta E \tag{93}$$

where U_{rep} is the repulsive part of the Lennard-Jones (6–12) interaction potential, ΔE an average excitation energy and f a weighting factor ($f = {}^1/_2$ for like atoms). Then the paramagnetic shielding effect is calculated (the diamagnetic $\sigma^{(1)}$ is negligible by comparison) resulting in

$$\Delta \sigma_{\text{rep}}^{(2)} (\text{Xe}) \cong \frac{2\, e^2 h^2 f N r_0^3 \epsilon \mathscr{H}_{12}(y)}{9\, \pi m^2 c^2\, (\Delta E)^2 y^2} \left\langle \frac{1}{r^3} \right\rangle_{5p} \tag{94}$$

The $(\sigma_1)_{\text{rep}}$ data so calculated turned out to be about 25% of the $(\sigma_1)_{\text{w}}$ values, and although both are negative, their sum does not nearly add up to the experimental σ_1 for quite a few of the binary systems.

The authors indicate that the shortcoming may be due to an inadequate repulsive potential. They argue that the repulsion effect must be mainly between the Xe atom and the peripheral H and F atoms of the perturbing solvent (reminescent of Raynes' solvent site factor, as described in section 6.2). They adopt a two-parameter repulsive potential as follows

$$U_{\text{rep}} = A \exp(-\lambda r') \tag{95}$$

with A and λ parameters from various literature sources. The distance r' is from the centre of the Xe atom to peripheral H or F; however, rotation of the solvent was neglected and only *linear* configurations such as Xe . . . H–C were considered. Such potentials will be deeper and steeper than corresponding Lennard-Jones (6–12) potentials. Although the authors did not pursue this approach quantitatively, it did produce some qualitative successes such as the desired inversion in the relative magnitude of the σ_1 parameters for the Xe . . . CH_4 and Xe . . . CF_4 systems. Unfortunately, this approach does little to remove the discrepancy for the Xe . . .Xe system which discrepancy is by far the largest. It would be interesting to see whether the discrepancies noted (either expressed in σ_1 or B parameters) could be removed by Mohanty and Bernstein's empirical approach. The important conclusion of Jameson, Jameson and Gutowsky remains, however, namely that relatively small changes in the chosen repulsive potential can have drastic effects on the shielding.

This point is perhaps further corroborated by the calculations of Adrian [67] which actually preceded those of Jameson, Jameson and Gutowsky. Adrian's work is probably in error with regard to his conclusion that in a Xe. . . Xe system the dispersive part of σ_{w} would be negligibly small. His calculation of the repulsion contribution, however, is roughly parallel to that of Jameson *et al.*, although there are significant differences. One of these is Adrian's use of a Buckingham (6-exp) potential. The interesting point is that Adrian calculates a $(\sigma_1)_{\text{w}}$ (repulsion only) for the Xe. . . Xe system, which is about 66% of the observed value as compared to 11% in the JJG calculation. The former number would fit very well with the calculated finding (JJG) that in the Xe. . . Xe system the dispersive contribution is almost 40% of the observed σ_1; however, Jameson *et al.* claim [66] that Adrian's result is so high mainly because of an error of almost a factor three in the evaluation of the overlap integral.

In sections 9.2, 9.6, 10.1 and 14.3 we will return to various other aspects of ^{129}Xe σ_{w}.

7.4. Pressure Effects on σ_w

A completely different approach to "repulsion" was given by Buckingham and Lawley [68], who calculated the shielding of an H atom under a uniform compression. They did this by using the wave functions of an H-atom at the centre of an impenetrable sphere as developed by Michels *et al.* [69]. Using Ramsey's formulation for the diamagnetic screening the authors then find that the shielding should *increase* due to such a cage. The results are

$$\psi = \exp\{-r'/(1+\beta)\}\ [1-\beta\sum_{s=1}^{\infty} b_s\, r'^{(s-1)}] \tag{96}$$

for $0 < r' < r'_0,\ r' = r/a_0$ and $r'_0 = r_1/a_0$,

$$\beta = \left[\sum_{s=1}^{\infty} b^s r'_0{}^{(s-1)}\right]^{-1} \text{ and } b_s = 2^{s-1}/s!\,(s-1)!$$

$$\sigma^d = \frac{e^2}{3\,mc^2 a_0}\ \frac{\int_0^{r'_0} \exp\{-2r'/(1+\beta)\}\left[1-2\,\beta)\sum_{s=1}^{\infty} b_s r'^{(s-1)}\right] r'\,dr'}{\int_0^{r'_0} \exp\{-2r'/(1+\beta)\}\left[1-2\,\beta\sum_{s=1}^{\infty} b_s r'^{(s-1)}\right] r'^2\,dr'} \tag{97}$$

For a cage with a radius of 5 Bohr radii (corresponding roughly to a pressure of 5,000 atm) they find $\Delta\sigma/\sigma = +1.42\%$. In liquids, where the internal pressure usually exceeds 1000 atm an upfield effect of several tenth of a ppm should therefore be expected for protons. However, this is contrary to experimental findings. In fact the gas-to-liquid shift for non-polar, isotropic molecules is always downfield and even stronger so (by a factor of nearly 2, see section 10.2) then predicted on the basis of extrapolation from the density dependence in the low pressure gas range. Apparently other effects obscure the picture and prevent the observation of this pressure effect. However, as Buckingham and Lawley write; "at sufficiently high pressure the cage effect, or the more general effect of non-bonded electron-electron repulsions it represents, presumably dominates and a positive value for $\partial\sigma/\partial p$ should be observed".

Such a study of $\partial\sigma/\partial p$ should be very rewarding. We don't know of any such published study but we would like to point out that the required high pressure techniques in liquids are apparently rather simple. We are referring here to the work of Yamada [70–72] and of Von Jouanne [73, 74], who, using glass or quartzglass tubes in an unmodified A60A have measured liquids under up to 2700 atm of pressure.

Chapter 8. The Effects of Higher Order Dispersion Terms

We have already mentioned previously (section 6.2) that Raynes' treatment of a solvent site factor is, at least formally, equivalent to the inclusion of higher order dispersion terms. It should be realised, however, that even solvent *atoms* have higher order dispersion terms. Mohanty and Bernstein [52] have remarked that they tried to incorporate these effects but that they found no improvement. No details were given, but one may presume that only a new set of empirical B values emerged with the same sort of statistical scatter.

It is not too difficult to project the effects of higher order dispersion terms on σ_w as the following derivations will show. For a pair of interacting molecules one may write

$$\phi^{\text{disp}} = -\frac{C}{r^6} - \frac{C'}{r^8} - \frac{C''}{r^{10}} \tag{98}$$

Margenau [75] has calculated the coefficients C, C' and C'' for a large number of molecules and atoms, starting from the semi-classical Drude theory for the dispersion of light. From his calculations it would appear that the r^{-10} term never contributes more than 2% to the total ϕ^{disp}; hence this term will be ignored. He also found, however, that the dimensionless parameter C'/Cr_0^2 is typically between 0.17 and 0.23; in other words the r^{-8} term would contribute roughly 20% to the total dispersion energy. The constant C might be identified with $4\,\epsilon r_0^6$ of the Lennard-Jones formulation; it then would follow that $C' \simeq 4\,\epsilon(0.2\,r_0^8)$. Before one can add these terms together, however, it is important to realise that the Lennard-Jones parameters have to be modified; these parameters are based normally on a (6–12) potential, but they will take on new values when determined using a (6,8–12) dependency.

Using asterisks to indicate these modified parameters one may now write

$$\phi = k\epsilon^* \left\{ -\left(\frac{r_0^*}{r}\right)^6 - 0.2\left(\frac{r_0^*}{r}\right)^8 + \left(\frac{r_0^*}{r}\right)^{12} \right\} \tag{99}$$

It can then be found from Eq. (99) that $0.967\,r_0^*$ represents the distance of zero potential, that ϵ^* is the depth of the well and (by differentiating) that $k = 2.93$. One can then say that Eq. (99) represents the same physical truth as the Lennard-Jones (6–12) potential of Eq. (42). In particular the repulsive terms should be the same; equating these gives

$$\epsilon^* = 1.365\,\epsilon \tag{100}$$

Equating next the dispersive parts of Eqs. (99) and (42) and solving this for an average distance $\bar{r} \approx 2^{1/6} r_0$ one finds

$$r_0^* = 0.977\,r_0 \tag{101}$$

Inclusion of the r^{-8} term will therefore in general give a deeper well, a shorter distance of zero potential ($0.967\, r_0^* = 0.944\, r_0$) and consequently a steeper rise into the repulsive range.

If one assumes that $\overline{F^2}$ remains proportional to the total dispersive energy, then, leaving out repulsion for the moment, the Eqs. (99, 40) and (44) lead to the following expression for σ_w;

$$\sigma_w{}^{6,8-12} = \frac{-\pi NB\epsilon\alpha_2 I_2}{\epsilon^*\, r_0^{*3}\, y^{*4}}\; \mathcal{H}_6(y^*) \left[1+0.2\, \frac{\mathcal{H}_8(y^*)}{\mathcal{H}_6(y^*)}\right] \tag{102}$$

The ratio $\mathcal{H}_8(y)/\mathcal{H}_6(y)$ is virtually independent of y (in the common $y = 1.5$ to 2.5 range) and can be found to be equal to 0.682 [41]. Using this and Eqs. (100, 101, 102) and (45) we now can express the effect of inclusion of the r^{-8} term as follows

$$\frac{\sigma_w{}^{6,8-12}}{\sigma_w{}^{6-12}} = 0.893\, \frac{\mathcal{H}_6(y^*)}{y^{*4}} \;\Big/\; \frac{\mathcal{H}_6(y)}{y^4} \tag{103}$$

The ratios $\mathcal{H}_6(y)/y^4$ are rather strongly dependent upon y. Since from Eq. (100) it follows that $y^* = 1.1684\, y$, the magnitude of the effect will depend on y. In Table 21 we have calculated the r^{-8} effect on σ_w for a number of y values.

Table 21. The calculated effect of an r^{-8} dispersion term on σ_w

y	$\mathcal{H}_6(y)/y^4$	$\mathcal{H}_6(y^*)/y^{*4}$	$\sigma_w{}^{(6,8-12)}/\sigma_w{}^{(6-12)}$
1.5	6.04	6.675	0.99
1.75	6.66	7.79	1.04
2.0	7.59	9.51	1.12
2.25	8.92	12.14	1.21

Since most σ_w measurements are done in the range $y = 1.8$ to 2.2 (room temperature, medium to large molecules) the effect is seen to be small and to range from about $+5\%$ to $+20\%$. Table 21 will in fact exaggerate somewhat the effect; at high y (larger ϵ or lower T) the average distance $\bar{r}$ will be more and more approaching the condition $\bar{r} = 2^{1/6} r_0$; at lower y a larger $\bar{r}$ should be used so that a larger coefficient in Eq. (101) will result which will reduce the $\sigma_w{}^{(6,8-12)}/\sigma_w{}^{(6-12)}$ ratio. It is unlikely, therefore, that the r^{-8} effect can be detected. Inclusion would only reduce the resulting B parameters somewhat. Varying y by changing solvent introduces the uncertainties in the various ϵ/k and r_0 data; this alone can easily account for $\pm\, 30\%$ variation in extracted B parameters. Changing temperature is not likely to reveal the effect either; In Rummens' work on the temperature effect in gaseous and liquid ethane [24] y ranged only from 1.55 to 1.80, which according to Table 21 would produce no more than a 5% change in σ_w due to the r^{-8} dispersion term.

It appears safe to conclude that for virtually all intents and purposes the higher order dispersion terms and their effects on σ_w may be ignored. This conclusion would not be altered if additionally a site factor and/or a repulsion factor were taken into account.

Chapter 9. The Parameters B

In this Chapter we shall summarise what is known of the parameter B. We will discuss the various quantum mechanical calculations for B and will also summarise the empirically obtained B parameters for non-polar molecules. The parameter B has only real significance in the formulations of $\sigma_w = -B\overline{F^2}$ (or $\sigma_w = -BE^2$). However, there are theories which calculate σ_w *directly*, without any intermediate electric field. In these cases it is still possible to equate such σ_w with $-B\overline{F^2}$ where some model for $\overline{F^2}$ has been inserted. With such a "simulated" field one then arrives at "simulated" B parameters. Such calculations have also been included in the present chapter. Therefore the material of this chapter will allow a comparison between the effective electric field theories and the direct interactive theories for σ_w.

9.1. Marshall and Pople's Calculation of B for an H atom [9]

Marshall and Pople have calculated the shielding of a single hydrogen atom that is simultaneously subjected to a uniform electric field E. They start by expanding the Hamiltonian into terms in each order in E and H (up to 2nd order). The total wave function is expanded likewise and the Schrödinger equations solved. This gives each component of the wavefunction in terms of polar coordinates, including the angle θ between E and H. The wavefunction corresponds to a certain current density distribution, which in turn causes an extra magnetic field ($= -\sigma H$) at the nucleus. The current density vector j is given by

$$j = \frac{-e^2}{2\,mc}(H \times r)\,\psi^*\psi - \frac{e\hbar}{2\,mi}(\psi^*\nabla^2\psi - \psi\nabla^2\psi^*) \tag{104}$$

where ψ is the (normalised) total wavefunction of the form

$$\psi = \psi_{00} + \psi_{01} + \psi_{10} + \psi_{11} + \psi_{20} + \ldots \tag{105}$$

(the suffix indicating the order of perturbation of E and H respectively). Once j has been found the field ΔH can be found

$$\Delta H = \int \frac{r \times j}{cr^3}\,d\tau \tag{106}$$

from which σ follows as $\sigma = -\Delta H/H$. ΔH and σ can be calculated in two parts according to the two terms in the expression for j. This first term may be called the "diamagnetic" term $\sigma^{(1)}$

$$\sigma^{(1)} = \frac{e^2}{2\,mc^2}\int \frac{\rho\sin^2\theta}{r}\,d\tau \tag{107}$$

Note that the problem is only *axially* symmetric because of E and that therefore Eq. (107) will not reduce to the Lamb formula and that $\sigma^{(1)}$ will be dependent on the angle between E and H. The results are

$$\sigma_\parallel^{(1)} = \frac{e^2}{3\,mc^2 a}\left[1 - \frac{439\,a^4 E^2}{40\,e^2}\right] \tag{108}$$

$$\sigma_\perp^{(1)} = \frac{e^2}{3\,mc^2 a}\left[1 - \frac{641}{80}\,\frac{a^4 E^2}{e^2}\right] \tag{109}$$

The second term in j gives a "paramagnetic" contribution $\sigma_\parallel^{(2)}$ equal to zero and a contribution $\sigma_\perp^{(2)}$ as follows;

$$\sigma_\perp^{(2)} = \frac{-233\,a^3 E^2}{144\,mc^2} \tag{110}$$

Note that the "paramagnetic" term $\sigma_\perp^{(2)}$ has the same sign as the "diamagnetic" terms and that $\sigma_\perp^{(2)}$ contributes 26% of the total $\Delta\sigma$ [see Eq. (111)]. It then follows that

$$\Delta\sigma = (2\,\Delta\sigma_\perp + \Delta\sigma_\parallel)/3 = \frac{-881\,a^3 E^2}{216\,mc^2} \tag{111}$$

so that $B = -\Delta\sigma/E^2$ is given by

$$B = \frac{881\,a^3}{216\,mc^2} = 0.738 \cdot 10^{-18}\ \text{esu} \tag{112}$$

9.2. Calculation of B for H Atoms and Rare-gas Atoms According to Jameson, Jameson and Gutowsky [66]

In this, approximative, approach the total wavefunction correct to the first order is given as

$$\psi^{(1)} = \psi^{(0)} + \Sigma\,\lambda_n^2\psi_n$$

The authors then show that λ is approximately given by

$$\lambda^2 = \Sigma \lambda_n^2 \approx \alpha E^2 / 2 \, \Delta E \tag{113}$$

where α is the polarisability, E is the perturbing electric field and ΔE is an average excitation energy. They then show that the paramagnetic contribution can be expressed in terms of orbital populations and through these in terms of λ^2. The change in $\sigma^{(2)}$ due to each considered excitation is then expressable in ΔE (of that particular transition), in $<r^{-3}>$ for each orbital involved (obtainable from atomic spectra) and in E^2. Division of $\Delta \sigma^{(2)}$ by $-E^2$ then gives contributions to the B parameter due to the "paramagnetic" part. For each excitation considered, a somewhat different B value may follow, but the variation is not large and a simple average may be used as a "best" value.

Where appropriate a diamagnetic contribution $\Delta \sigma^{(1)}$ may also be calculated but this involves terms like

$$\Delta \sigma^{(1)} = \frac{e^2 \lambda^2}{3 \, mc^2} \left(<n \mid \frac{1}{r} \mid n> - <0 \mid \frac{1}{r} \mid 0> \right) \tag{114}$$

which are more difficult to evaluate or to estimate. For heavy atoms (like ^{129}Xe) the paramagnetic term $\Delta \sigma^{(2)}$ is very large, up to two orders of magnitude larger than $\Delta \sigma^{(1)}$, so that the latter may be ignored.

Our impression is that B values so obtained (see Table 22) may be quite good for heavy atoms, but the approximations inherent in Eqs. (113) and (114) would make the procedure less reliable for light atoms. The value $B(\text{H}) = 0.35 \cdot 10^{-18}$ esu obtained in this manner (only using the $1s \rightarrow 1p$ excitation) compares poorly with $B(\text{H}) = 0.74 \cdot 10^{-18}$ esu obtained by Marshall and Pople. Also the value for ^{3}He, $B(\text{H}) = 0.075 \cdot 10^{-18}$ esu, is very low; one would expect a value *larger* than that from an H atom, although the reverse is not impossible.

Table 22. B parameters (in units of 10^{-18} esu)
as calculated by Jameson, Jameson and Gutowsky [66]

Probe	B
H	0.35
He	0.075
Ne	5.6
Ar	41.3
Kr	124.7
Xe	337.3

9.3. σ_W of Two Interacting H Atoms According to Marshall and Pople

Marshall and Pople [45] have given a perturbation theory which is second order in the electrostatic dipole interaction between the two atoms ($\lambda^2 \mathscr{H}_{200}$), in the interaction of the electron with the magnetic field of the perturbing atom ($\mu^2 \mathcal{H}_{020}$) and in the electronic interaction with the nuclear moment ($\nu^2 \mathcal{H}_{002}$). The energy of the lowest state can then be written as

$$\epsilon = \lambda^2 \epsilon_{200} + \lambda^2 \mu^2 \epsilon_{220} + \lambda^2 \mu\nu\epsilon_{211} + \ldots \tag{115}$$

Other terms were also considered, but they are either zero, such as $\lambda\mu^2$, $\lambda\mu$ and $\lambda\nu$, or their magnitude is negligible (such as $\lambda^2\nu^2$). After numerical evaluation of the energies ϵ_{ijk} using variational techniques, Eq. (115) can be developed in a power series in distance R, magnetic moment M and the magnetic field H_0; the shielding σ is then formed by the coefficient to the term $^1/_2 MH_0$. The results are

$$\sigma_{\parallel} = \sigma_0(1 + 3\,r^{-3} - 21.59\,r^{-6} + \ldots) \tag{116a}$$

$$\sigma_{\perp} = \sigma_0(1 - \frac{3}{2}r^{-3} - 25.00\,r^{-6} + \ldots) \tag{116b}$$

for the situations where the H atom pair is parallel and perpendicular to the z-axis respectively. Upon averaging over all orientations of the pair relative to H_0, Eqs. (116) result in

$$\sigma = \sigma_0(1 - 23.86\,r^{-6}) \tag{117}$$

and hence

$$(\sigma_{\text{w}})\text{pair} = -\frac{23.86\,\sigma_0}{r^6} \tag{118}$$

where σ_0 is the shielding of a single hydrogen atom. If one combines Eqs. (43) and (44) to obtain $(\sigma_{\text{w}})\text{pair} = -3\,B\alpha_2 I_2/r^6$, the combination with Eq. (118), using parameters for atomic hydrogen, then results in a value of $B = 0.20 \cdot 10^{-18}$ esu [44, 66]. This value is therefore considerably smaller than the value of $B = 0.74 \cdot 10^{-18}$ esu derived for a single H atom in a static electric field [Eq. (112)]. The difference may be due in part to the choice of the RBB expression for $\overline{F^2}$, [Eq. (44)], the latter being open to considerable criticism (see section 5.2). Nevertheless, this cannot be the only reason; the empirical proton B parameters, based on the RBB model with the same model for $\overline{F^2}$ but excluding repulsion effects (which were also ignored by Marshall and Pople) are between 0.5 and $1.0 \cdot 10^{-18}$ esu, depending on various other assumptions (see Table 25). On the other hand it is also plausible to expect that the oscillating dipole field of a perturbing atom at close range is not at all comparable with a static (or low frequency) external field that is considered uniform over the probe atom.

The variational techniques employed may have created another substantial error (see also next section).

9.4. The Perturbation Calculations of Yonemoto

Yonemoto [76] has carried out calculations on the shifts of a H atom perturbed by neighbouring atoms. His procedure runs largely parallel to that of Marshall and Pople, except that he calculates the perturbation terms without recourse to variational techniques. As a consequence his results are couched in terms of a mean excitation energy ΔE. For two interacting H atoms he finds

$$(\sigma_w)\text{pair} = -\sigma_0 \left(\frac{9\,a^4 e^2}{\Delta E^3} + \frac{4\,a^3 e^6}{\Delta E^3}\ r^{-6} \right) \tag{119}$$

the two terms in Eq. (119) corresponding, respectively, to the $\lambda^2 \mu^2$ and $\lambda^2 \mu \nu$ terms of Eq. (115). Numerically, Eq. (119) gives $\sigma_w = -0.30$ and -0.035 (not 0.025 as given in [76]) ppm at 4 a.u. and 3 Å respectively; the Marshall and Pople model of Eq. (118) giving $\sigma_w = -0.10$ and -0.012 ppm at these distances. Yonemoto's calculation therefore produces σ_w values, exactly three times larger than the Marshall and Pople method. Equating Eq. (119) again to $-B\overline{F^2}$ of the RBB model, this would then correspond to $B = 0.60 \cdot 10^{-18}$ esu. The above numerical results were obtained by using the value of the ionisation potential of hydrogen ($22 \cdot 10^{-12}$ erg) for ΔE, but this is consistent within the RBB model for $\overline{F^2}$. Note that Yonemoto's value of $B = 0.60 \cdot 10^{-18}$ esu is within the range of the empirically determined B values.

Yonemoto also calculates relative Van der Waals interactions (dispersion only) between an H atom and halogen atoms as given in Table 23.

Table 23. Relative σ_w for H. . X pairs, according to Yonemoto [76], at constant and equal interatomic distance

perturber	H	F	F$^-$	Cl	Cl$^-$	Br	Br$^-$
relative σ_w	1	3.1	3.6	10	12.5	14	18
relative $\alpha_2 I_2$	1	2.1	–	5.4	–	7.0	–

We have also indicated approximate $\alpha_2 I_2$ values for the perturbing atoms. It is seen that, if a simulated field model of the RBB type is used, one can no longer maintain a B parameter which is independent of the perturber; between H. . .H and H. . .Br a factor of two difference develops.

Yonemoto also has considered hydrogen atoms *bonded* in a molecule. As a first approximation he takes such atoms as being "polarised" i.e. that a certain amount of p_z character is mixed in with the 1s orbital, if z is the direction of the bond. This results in an expression for σ_w with a proportionality factor $(1-1.4\,C^2)$ where C is the mixing coefficient. Yonemoto finds that for $C = 0.3$ his theory reproduces the A and B parameters of the Buckingham model for the shielding effect in an electric field [77] ($\sigma_E = -AE_z - BE^2$). Therefore, if a hydrogen in a bond can be described as a polarised atom, the Van der Waals shielding is reduced by about 13%, which would result in a B parameter, reduced by the same amount ($B = 0.52 \cdot 10^{-18}$ esu). Finally, Yonemoto undertook to calculate the B parameter for a real H_2 molecule. A simple localised LCAO-MO model with Slater type atomic orbitals was used. As far as σ_w is concerned, relative to an isotropic perturber the effect is again a reduction of σ_w or an equivalent reduction of B. A conversion factor of about 0.75 pertains to obtain $\sigma_w(H_2 \ldots X)$ from $\sigma_w(H \ldots X)$. It appears therefore that the B parameter of a bonded H atom is somewhat less than that for a free H atom, the degree of difference dependent on the degree of "polarisation" of the X–H bond. This is in accordance with the empirically found sequence $B = 1.06$ for hydrocarbon C–H bonds [15], $B = 0.84$ for CHF_3 [42] and $B = 0.38 \cdot 10^{-18}$ esu for Cl–H [15]. In any case it would appear that the B parameter for C–H in hydrocarbons is constant within perhaps a few percent irrespective of the bonding state of the carbon atom.

9.5. Kromhout and Linder's Calculation

Kromhout and Linder noted [44] that it is not necessarily correct to use a B parameter relating to a static field to a phenomenon which really rests on rapidly fluctuating fields. They pointed out that if the expression $\sigma_w(\text{pair}) = -B\overline{F^2} = -3 B\alpha_2 I_2/r^6$ [see Eqs. (43) and (44)] is equated to $\sigma_w(\text{pair})$ of two interacting H atoms as calculated by Marshall and Pople [45] a value of $B = 0.20 \cdot 10^{-18}$ esu will result as compared to $B = 0.74 \cdot 10^{-18}$ for an H atom in a static field.

Their calculation starts with perturbation treatment of the Hamiltonian of two atoms in a magnetic field, interacting through a dipole-dipole term.

$$\mathcal{H} = \mathcal{H}_{000} + \lambda \mathcal{H}_{100} + \mu \mathcal{H}_{010} + \nu \mathcal{H}_{001} + \mu\nu \mathcal{H}_{011} \tag{120}$$

with wavefunctions and energies expressed in power series of λ, ν and μ

$$\epsilon = \sum_m \sum_n \sum_q \epsilon_{mnq} \, \lambda^m \mu^n \nu^q \tag{121}$$

Only the term ϵ_{211} is considered, i.e. the term proportional to the field, to the nuclear moment and to the square of the dipole-dipole operator. This is one fundamental difference with the calculations of Marshall and Pople and of Yonemoto who also included the $\lambda^2 \mu^2 \epsilon_{220}$ form.

The results have already been given as Eqs. (54) and (57). We can now extract expressions for B parameters. Combination of the Raynes, Buckingham and Bernstein expression for $\sigma_w(\text{pair})$ with Eq. (57) gives

$$B_{\text{RBB}} = \frac{\sigma_0 \alpha_1 C (3 I_1 + 2 I_2)}{2 (I_1 + I_2)^2} \tag{122}$$

The most interesting aspect of Eq. (122) is that it shows B to be a function, not only of the solute, but also of the solvent. However, it must be remembered that this is only so within the framework of the employed effective $\overline{F^2}$ (or "simulated field" to use the terminology of Kromhout and Linder). If for example, the Howard, Linder and Emerson field [Eq. (18)] is used, then it follows that

$$B_{HLE} = \frac{\sigma_0 \alpha_1 C (3 I_1 + 2 I_2)}{I_1 (I_1 + I_2)} \tag{123}$$

It is interesting to note that for $I_1 = I_2$ it turns out that $B_{HLE} = 4 B_{RBB}$, a difference which no doubt originates in Eq. (16) as compared to its macroscopic dielectric analogon.

It might be thought that the I_2 dependence of B [Eq. (122)] may have been the reason for the variation in empirical B values for ^{129}Xe and ^{19}F measurements [66, 52] using the RBB method. We have found, however, that the empirical B's show only a crude correlation with $(3 I_1 + 2 I_2)/(I_1 + I_2)^2$. If such a dependence for B were built in, only a partial improvement would be obtained; the range of B values would decrease by about 30%. Neither can the discrepancy between the two lines of Table 23 be dissolved by the use of Eq. (122) or Eq. (123). Eq. (122) can be used to calculate B parameters. Using the molecular parameters as given by Kromhout and Linder, the results are as per Table 24.

Table 24. Effective B parameters calculated from Eq. (122) for X . . . X interactions ($I_1 = I_2$)

	$B_{RBB} \cdot 10^{18}$ (esu)
H. . .H	0.21
He. . . He	0.17
Ne. . . Ne	4.1
Kr. . . Kr	252
Xe. . . Xe	914
CH_4 . . CH_4	0.59
CF_4 . . CF_4	18

A comparison with Table 22 reveals large differences; perhaps an indication of the difficulty of these calculations. The difference between the B parameters for H and for CH_4 is striking. This is only in a small part (15%) due to the theory of Kromhout and Linder and then only because σ_0 is different. A greater part arises because the authors take the *atomic* values for α_1 and I_1. One must remember, however, that the difference arises in a large part because of a particular choice of simulated field $\overline{F^2}$ model. Nevertheless, the prediction that B (CH_4) would be larger than B(H) is in striking contrast with the calculations of Yonemoto.

9.6. Empirical Determinations of B for Atoms and Non-polar Molecules

Rummens has discussed the various sets of B parameters for CH that will arise depending on the model used [24]. A summary, based mostly on the experimental data of Raynes, Buckingham and Bernstein [15] and of Mohanty and Bernstein [52] is given in Table 25.

This table, better than anything else, demonstrates the uncertainties and dilemma's regarding the B parameter. It was originally thought [15] that the variation in B for pure CH_4, C_2H_6 and C_2H_4 was due to large experimental uncertainties. Later experiments [52], showed however, that the original differences were genuine. It is interesting to see that after inclusion of a site correction, after recalculation using a consistent set of potential parameters and after a correction of σ_a (mainly for C_2H_4), an almost constant value for B emerges, independent of the solute. Since for these pure hydrocarbons the ionisation potentials are almost the same, this finding is consistent with Kromhout and Linder's theory [Eq. (122)] and also with the calculation of Yonemoto on bonded hydrogen atoms (section 9.4).

Table 25. Empirical B parameters for ^{1}H in C–H bonds, obtained by the method of Raynes, Buckingham and Bernstein and variations thereof (B in units of 10^{-18} esu)

CH_4	C_2H_6	C_2H_4	Average and rel. stand. dev.	Ref.	Remarks
0.55	1.44	1.19	1.06 ±43%	15	No correction for site, repulsion or σ_a. For ethane ϵ/k and r_0 from viscosity, others from $B(T)$.
0.76	1.74	1.52	1.34 ±38%	16	As above, but with site and repulsion correction [See Eq. (87)].
0.41	0.92	0.82.	0.72 ±37%	16	As above, but site correction only [See Eq. (76)].
0.57	0.59 (0.60)	0.82	0.66 ±21%	24	As above, but with $<(\sigma_1)_w>$ = 7.8 (see Table 12) for CH_4 and with all ϵ/k and r_0 from $B(T)$.
0.57	0.55	0.50	0.54 ±5%	24	As above, but with estimated correction for σ_a.
0.53±0.18 in CH_4 0.73±0.18 in SF_6 0.74±0.17 in CF_4 0.94±0.16 in Kr 1.55±0.22 in SiF_4 2.10±0.14 in Xe			1.09 ±55%	52	No corrections for site or repulsion

Missing from Table 25 is the value of $B = 1.0 \cdot 10^{-18}$ often ascribed to Buckingham. Buckingham [77] did end up with the much quoted expression $\sigma = \sigma_0 - 2 \cdot 10^{-12} E_z - 1.0 \cdot 10^{-18} E^2$. However, since he used Marshall and Pople's results for the H atom in a field E, the B value is of necessity identical to $0.74 \cdot 10^{-18}$. Presumably the factor 1.0 is just the result of rounding off.

In Table 26 the ^{19}F results are tabulated.

Table 26. Empirical B parameters for ^{19}F in non-polar isotropic systems (in units of 10^{-18} esu)

CF$_4$	SiF$_4$	SF$_6$	solvent	Ref.	Remarks
22.6±0.9	31.3±1.0	38.2±2.2	CF$_4$		Measurements at 30 °C.
26.6±1.1	43.1±2.4	44.6±2.2	SiF$_4$		RBB method [Eq. (45)]
29.7±1.5	40.6±2.0	45.2±1.9	SF$_6$	52	No site or repulsion correction
28.4±1.0	46.7±1.8	53.7±3.0	CH$_4$		
28.3±0.9	43.0±1.0	48.8±2.7	Kr		
43.0±1.5	63.0±1.6	65.0±2.1	Xe		
262 ±20	690 ±47	782 ±67	see above	52	Measurements at variable temperature. No site factor correction. Empirical repulsion factor included [see Eqs. (90) and (91) and Table 20].
16.4±13%	43.5±12%	29.5±8%	CF$_4$, SiF$_4$, SF$_6$, CH$_4$, CHF$_3$	42	No site or repulsion correction
24.1±8%	59.2±7%	40.0±5%	as above	16	Date from [42] but site and repulsion correction included [Eq. (87)].

It is clear that the ^{19}F B parameters for CH$_4$, SiF$_4$ and SF$_6$ are significantly different. Petrakis and Bernstein have argued, however, that this is to be expected [42]. They claim that in the series CF$_4$, SF$_6$, SiF$_4$ there is an increasing double bond character. This is turn means increased electron density at the ^{19}F nucleus and therefore larger shielding. Indeed in the original work of Petrakis and Bernstein [42], the B parameters increase in that direction. However, if one takes the most recent, and presumably more accurate, data of Mohanty and Bernstein the B parameters increase in the direction CF$_4$, SiF$_4$, SF$_6$. This may also be taken as one further criticism against the repulsion contribution employed (see section 7.2). Without this repulsion model there is a significant variation of B for each solute as a function of solvent gas. Again the spread is too large to be accounted for by Kromhout and Linder's theory [Eqs. (122) or (123)]. All empirical B values of Table 26 are one or two orders of magnitude larger than the B value for Ne ($B = 5.6 \cdot 10^{-18}$ esu, Table 22).

We finally give in Table 27 the empirical B parameters for ^{129}Xe.

Table 27. Empirical B parameters for ^{129}Xe as obtained by
Jameson, Jameson and Gutowsky [66] using the RBB method [15]

solvent	$B \cdot 10^{18}$ (esu)
CH_3F	318
CF_4	369
CHF_3	375
Ar	382
CH_2F_2	395
CO_2	413
Kr	506
CH_4	592
HCl	606
Xe	837

Polar solvents have been included in Table 27 contrary to most other data in this review, simply because with ^{129}Xe the polar contributions BE^2 are quite small compared to BF^2 and never amount to more than 5%. (However, according to Kromhout and Linder's [44] and Yonemoto's [76] theories the B's in BE^2 and BF^2 are fundamentally and substantially different!).

Chapter 10. σ_{w} in Dense Media

10.1. The Effects of Higher Order Collisions in Gases

Whereas the macroscopic continuum theories for medium effects (Chapter 2) pretend to be universal in their applicability to all conditions of temperature, pressure and density in all phases, we shall see that the experimental evidence is contrary to this. With the statistical mechanical model of Raynes, Buckingham and Bernstein [15] there is a clear alternative. While at low densities the σ_{w} effect can be accounted for by binary collisions, the virial theorem of Eq. (37) leaves the possibility of adding terms to account for ternary and higher order collisions. Such an approach seems reasonable at least for gases at high density. According to the virial theorem one would therefore expect that at high density the dependence of the medium shift on density would become non-linear. An interesting example of such a non-linear shift has been found with ^{129}Xe. Initially a linear behaviour was found with densities up to 300 amagats. Streever and Carr [78], and Hunt and Carr [79] report $\sigma_1 = -0.43$ ppm/amagat at 25 °C. More recently, working with greater precision, Kanegsberg, Pass and Carr [80] found a distinct non-linearity, requiring a third virial term σ_2. They also found a distinct temperature dependence for σ_1 and σ_2. Their results are $\sigma_1 = -0.61, -0.54, -0.51$ and -0.49 ppm/amagat at 20, 40, 60, and 80 °C respectively ($\sigma_1 = -0.695 + 4.8 \cdot 10^{-3} t - 2.8 \cdot 10^{-5} t^2$), and σ_2 = 4.7, 2.6, 1.3 and $0.6 \cdot 10^{-4}$ ppm/amagat^2 at these same temperatures ($\sigma_2 = +7.31 - 0.15 t +0.78 \cdot 10^{-3} t^2$). Jameson, Jameson and Gutowsky [66] observed that actually the shift is only linear up to 100 amagats. Using only these data points they found $\sigma_1 = -0.548 \pm 0.004$ ppm/amagat (at 25 °C). However, upon including data points up to 250 amagats, they found that not only a quadratic but also a cubic term was required to fit the data within the precision limits of their data. The results are $\sigma_1 = -0.548 \pm 0.004$ ppm/amagat, $\sigma_2 = (-0.169 \pm 0.02) \cdot 10^{-3}$ ppm/amagat^2 and $\sigma_3 = (+0.163 \pm 0.01) \cdot 10^{-5}$ ppm/amagat^3. One may note that σ_2 is found to be negative. However, if σ_3 is assumed to be zero, a reasonable fit can still be obtained with $\sigma_1 = -0.55$ ppm/amagat and $\sigma_2 = +2.3 \cdot 10^{-4}$ ppm/amagat^2 in reasonable agreement with the findings of Kanegsberg, Pass and Carr. In any event the curvature is such that at high pressure a lower σ_w is found than anticipated from low-pressure measurement extrapolation.

Rummens has studied ethane in the gaseous (up to about 160 amagat) and the liquid state [24] and observed considerable curvature in σ vs ρ plots. His results could be condensed as follows:

$$\sigma_1 = -18.5 + 0.042 \, t \text{ (ppm cm}^3 \text{ mole}^{-1}) \tag{124}$$

$$\sigma_2 = +572 + 5.88 \, t + 0.105 \, t^2 \text{ (ppm cm}^6 \text{ mole}^{-2}) \tag{125}$$

Note that again a positive σ_2 is observed. Rummens points out that there is a great similarity between the theoretical expressions for $(\sigma_2)_w$ (second virial coefficient for Van der Waals screening) and the second virial coefficient $B(T)$ for the equation of state. Using tabulated B^* (T^*) data and ethane molecular parameters he calculates $(\sigma_1)_w = -15.3 + 0.066t$ ppm cm^3 mole^{-1} which is in reasonable agreement with the empirical finding of Eq. (124). Theorising that a similar resemblance between the third virial coefficients of screening and of the equation of state should then also exist, Rummens predicts that σ_2 should be positive and should have a quadratic temperature dependence. Again the ^{129}Xe and the ethane experimental data seem to bear out this prediction. (see section 11.3 for a further discussion of this point).

The above results are in strong contrast to those of Oldenziel [50] who measured CH_4 and C_2H_4 up to 600 amagat. In the region of 0–350 amagat the relation with ρ was found to be virtually linear but from 350–580 amagat there is a distinct curvature, both for CH_4 and C_2H_4, but towards *stronger* deshielding (i.e. opposite to the direction of curvature for Xe and C_2H_6. In Table 28 we reproduce the virial coefficients for CH_4 as determined by Oldenziel with a polynomial least squares method

$$\sigma - \sigma_0 = \sigma_1 \frac{\rho}{M} + \sigma_2 \left(\frac{\rho}{M}\right)^2 + \sigma_3 \left(\frac{\rho}{M}\right)^3 + \sigma_4 \left(\frac{\rho}{M}\right)^4 \tag{126}$$

71

Table 28. Best virial coefficients for CH_4 for varying length of virial expansion [Eq. (126)]

density (amagat)	number of terms	σ_1	σ_2	σ_3	σ_4
0–350	1	−7.7			
0–350	2	−6.8	−1.0		
0–350	3	−8.9	+4.2	− 2.9	
0–580	4	−4.8	−8.3	+10.3	−4.4

Oldenziel has also noted that at high density the exponential Boltzmann function in the expression for σ_w [Eq. (40)] is not generally correct, but must be replaced by some distribution function which has an oscillatory radial behaviour. In general such functions are not available, but for CH_4 it exists in numerical form [81]. Using this radial distribution function, Oldenziel finds that up to 100 amagat and again around 400 amagat, the result is exactly the same as with the Boltzmann factor, using binary collisions only. Around 250 amagat he calculates a σ_w about 4% (i.e. 0.03 ppm) less than with the exponential function; in other words if the measurements are restricted to 250 amagat, a *positive* σ_2 is predicted, which actually corresponds with the experimental result and with the calculated negative σ_2 over the 0–350 amagat region. Unfortunately no pair distribution data above 400 amagat are available.

10.2. The Gas-Liquid Transition

Some early studies, such as by Gordon and Dailey [23] on CH_4, C_2H_4, and by Streever and Carr [78] on Xe indicated that the density dependence was linear including points in the liquid region. This is in apparent disagreement with the results discussed in the previous section and with the finding (protons only [17]) that if one uses σ_1 or B parameters obtained from low density gas studies to calculate σ_w shifts in liquids one invariably calculates a σ_w which is approximately a factor 1.65 too small. In part this disagreement is due to poor experimental precision in those early studies. If one looks at Rummens' results on gaseous and liquid C_2H_6 [24], it is seen that the relation with density is far from linear; if one forces a straight line through the data, the maximum deviation is 2 Hz, which is just about equal to the precision of the Gordon and Dailey experiments. As noted above the zero-pressure-gas-to-liquid σ_w shifts are usually nearly twice as large as would be predicted by extrapolation from low density measurements to the liquid density. Rummens, Raynes and Bernstein [17] propose a scale factor $K_6^g = 1.46 + 0.0064\,t$ to be incorporated into Eq. (45) or Eq. (76). Liquid ethane is the one exception for which the σ_w of the liquid (in the $\rho = 0.3$ to 0.5 region) is *less* than calculated on a low pressure σ_1 basis. A scale factor larger than unity seemed at variance with the results at medium pressures which always produced a positive σ_2, *i.e.* a diminishing σ_w. It now appears likely that this apparent divergence is due to the upper limit of about 350 amagat of these medium pressure studies. The very high pressure data of Oldenziel show that eventually σ_w increases stronger than linearly with the density. It then follows that probably the σ_w *vs* ρ function is smooth, without a sharp discontinuity at the gas-liquid transition.

It might well be argued here that the virial theorem and its breakdown in binary, ternary and higher order collisions is not valid at all for liquids. One should perhaps recognise some form of cluster theory or in other ways reconcile the experimental fact of radial mass distribution functions. Or one might embark on Monte Carlo or Molecular Dynamics calculations for real liquids. But all of the above seems at present rather prohibitive in terms of computer size and time and in view of the very limited applicability to the simplest of liquids. In fact the observation of Linder and Hoernschemeyer [46] that the radial distribution function $\rho(r)$ plotted against r/r_0 appears universally identical, at least for those few liquids for which $\rho(r)$ is known, resulting in a universal constant $2/3\,r_0^3$ for the corresponding spatial integral [see Eqs. (58–61)] seems a hopeful sign that simplistic models may provide correct results.

It seems to us that the apparent universality of the constant K_6^g or of the radial distribution integral contains a special message; namely that *as far as NMR shifts are concerned*, the effect of a real liquid, surrounding a solute molecule, appears proportional to the sum of all binary interactions. Such an extra liquid effect could be due to the high internal pressure of liquids. But, as Buckingham and Lawley have shown [68], such pressure effect (i.e. caging-in of the wave functions) is rather small (estimated to about 0.005 ppm for a typical liquid, see section 7.4) and furthermore upfield, rather than the sought-for extra downfield effect. From the foregoing it seems more likely that there exists an effective pair potential which could incorporate gases as well as liquids.

There is one more consideration to be taken into account here, namely the *size* of the molecules, these having been small for the purpose of calibrating the parameter B, and being much larger in the gas-to-liquid studies. Larger molecules require a Kihara type of potential [Eq. (80)]; as the prelimenary results of Rummens and De Meyer have shown [62b], the introduction of such a potential increases the effectiveness and universality of the binary collision model.

Further information on the nature of K_6^g might be sought by studying the temperature dependence of σ_w and K_6^g as will be discussed in section 11.3.

Rests us to conclude at this stage that there is no readily available answer to the apparent discrepancy between gases and liquids. This should, however, keep no one from enjoying the unexpected but extremely useful universality of K_6^g; the above discussion holds the promise that it is not a mere fudge factor!

Chapter 11. The Temperature Dependence of σ_w

11.1. The Intramolecular Temperature Effect

Of all the possible aspects of σ_w and its changes, the dependence upon temperature takes a special place. It is the only effect which can (and does) also effect the shift of an isolated molecule. Since σ_w is defined as $\sigma_w = \sigma_{liq} - \sigma_0 - \sigma_b$, not only must the bulk susceptibility term be calculated at the appropriate temperatures, but in addition $d\,\sigma_0/dt$ must be considered. There is a special difficulty here, since until now it has been impossible to measure accurate *absolute* temperature dependencies. Virtually all known experimental data are relative to some chosen molecule so that one measures only the difference of two temperature dependencies. (For apparent exceptions see the end of this section) The first paper on the subject is that by Petrakis and Sederholm [82], who measured $d\sigma_w/dt$ for eleven gases relative to internal CH_4. Relative to this standard they found negative $d\sigma_w/dt$ values (except for C_2H_4) of magnitudes varying from zero (HBr and C_2H_6) to about -0.02 ppm/100 °C (for $SiMe_4$). Buckingham [83] has commented on these data; he pointed out that Petrakis and Sederholm's data refer to samples at 5–10 atm pressure, and that under these circumstances the intermolecular temperature effects are in the same order of magnitude as those reported by Petrakis and Sederholm. Petrakis and Sederholm tried to explain their results in terms of vibrational excitation. They indicated that stretching modes can never contribute much, (explaining the zero-effect for HBr) and that only certain low frequency torsonial modes could possibly be responsible. Even at that, their Δ_{eff} (i.e. NMR frequency difference for $v = 0$ and $v = 1$) parameters, required to fit the experimental data seem much too large, by about an order of magnitude. Buckingham [83] has also criticised this approach and has pointed out that rotational excitation (and subsequent centrifugal stretching) is a much more likely cause for intramolecular temperature dependence of chemical shifts. He estimated that even for a vibrational frequency as low as 500 cm^{-1} the rotational contribution is about equal to the vibrational contribution (at 300 °K). Rummens [24] has also commented on the data of Petrakis and Sederholm, pointing to an apparent equality of their intramolecular $d\sigma_0/dt$ data and the $d\sigma_w/dt$ data for the liquid state as given by Rummens, Raynes and Bernstein [17]. Actually this comment is in error; upon re-reading the Petrakis and Sederholm paper, the author has found that their chemical shift parameters δ are actually shielding parameters σ. The oversight of this caused an error in sign in the data quoted [24].

More recently Raynes *et al.* [84] have carried out very exact calculations on H_2 (and also HD, D_2, HT, DT and T_2). They find that for *ortho*-H_2 the temperature effect on σ is entirely due to centrifugal stretching and in fact mostly to the $v = 0$, $J = 1 \rightarrow v = 0, J = 3$ transition. The effect is calculated to be about -0.013 ppm/ 100 °C in the NMR accessible range of -100 °C to $+200$ °C.

Raza [55] and Raynes and Raza [59] have measured a number of gases as a function of temperature, relative to internal ethane. Since they used low partial pressures (2 atm for C_2H_6 and 1 atm or less for the solute) the contributions of the temperature dependence of the *inter*molecular effects were minimised.

The experimental data of both Petrakis and Sederholm, and of Raza and Raynes are given in Table 29.

The precision of both sets of data is probably $\pm$ 0.003 ppm/100 °C. Upon comparing the two sets it is apparent that the intermolecular contribution can be significant. Even the 2–3 atm, used by Raynes and Raza, may have been too high a pressure.

As Buckingham has shown [83], the leading rotational term in $d\sigma_0/dt$ is proportional to B_e/ω_e^2. For HBr this factor is much smaller than for H_2. The dependence

Table 29. Intramolecular $\Delta\sigma_0/\Delta t$ (ppm/100 °C) relative to ethane. According to Petrakis and Sederholm [82] and to Raza and Raynes [55, 59]

	ref [82] 6–10 atm	ref [55, 59] 2–3 atm
CH_4	0.000	0.000
HBr	0.000	
C_2H_2	−0.015	
C_2H_4	+0.006	
C_3H_6	−0.013	
C_4H_8	−0.012	
C_5H_{10}	−0.006	−0.006
C_6H_{12}	−	−0.005
$C(CH_3)_4$	−0.010	
$Si(CH_3)_4$	−0.021	−0.007
$Sn(CH_3)_4$	−0.017	−0.027
$CH_3C{\equiv}CCH_3$	−0.008	
C_6H_6		+0.022
$p-C_6H_4(CH_3)_2$		+0.020
$p-C_6H_4(CH_3)_2$		+0.047
$1, 3, 5\text{-}C_6H_3(CH_3)_3$		−0.030
$1, 3, 5\text{-}C_6H_3(CH_3)_3$		−0.030
$Si(OCH_3)_4$		0.000
$Si(OCH_2CH_3)_4$		+0.005
$Si(OCH_2CH_3)_4$		+0.025
$C_6H_4F_2$		−0.022

of σ_0 upon intermolecular distance is also different for HBr than for H_2, but this is likely to be a smaller factor. It would appear therefore that $d\sigma_0/dt$ for HBr might well be smaller than for H_2. This would mean that the relative $\Delta\sigma_0/\Delta t$ values of Table 29 are close to the absolute values. On this presumption it would then follow that the experimental data (Table 29) can never be explained by centrifugal stretching alone. Particularly for the larger molecules an increasing vibrational contribution (often of opposite sign) appears indicated.

The data of Table 29 are also important from a purely practical point. Molecules like 1, 3, 5-trimethylbenzene can only be measured in the gas phase at temperatures about 150 °C above room temperature. If such data are compared to infinite dilution data at room temperature, the resulting gas-to-liquid shift will be too small by 0.045 ppm (2.7 Hz at 60 MHz) for 1, 3, 5-trimethylbenzene, or too large by 0.070 ppm for the aromatic protons of *para*-xylene.

On the other hand, Louman [57] has considered that the $\Delta\sigma_0/\Delta t$ effect is very small if at all detectable. His data are relative to internal ethane in gas mixtures of 5 atm each for solute and reference (see Table 30). He therefore resorted to measure all gas shifts, required for gas-to-liquid shifts, at 180 °C. However, he did *not* measure $\Delta\sigma_0/\Delta t$ for benzene, *para*-xylene and mesitylene, for which molecules Raza has found rather large $\Delta\sigma_0/\Delta t$ values.

Recently Jameson and Jameson [85] have indicated how, at least in principle, the *absolute* intramolecular temperature dependence of σ_w can be measured. Their

Table 30. $\Delta\sigma_0/\Delta t$ (ppm/100 °C) relative to internal ethane at partial pressures of about 5 atm according to Louman [57]

	$\Delta\sigma_0/\Delta t$ ppm/100 °C
$CH_3C\equiv CCH_3$	0.000
$CH_3CH_2C\equiv CCH_2CH_3 \begin{cases} CH_3 \\ CH_2 \end{cases}$	−0.012 −0.012
trans-butene-2 $\begin{cases} CH_3 \\ CH \end{cases}$	−0.003 −0.011
iso-butane $\begin{cases} CH_3 \\ CH \end{cases}$	−0.005 +0.004
$CH_3CH=CHCH_3 \begin{cases} CH_3 \\ CH \end{cases}$	+0.005 +0.007
$(CH_3)_2C=C=C(CH_3)_2$	0.000

method is based on the fact that isolated atoms such as ^{129}Xe do not have an intrinsic temperature dependence of σ_0, because of the absence of rotational and vibrational states. Therefore, if at a number of Xe pressures the chemical shift difference between the ^{129}Xe signal and the signal of another (external) material is measured as a function of temperature and if the resulting $d\sigma/dt$ data are then extrapolated to zero Xe density, the resulting $(d\sigma/dt)_{\rho\,Xe=0}$ will be an absolute temperature dependence. The authors measured such absolute temperature dependencies for three pure liquids *viz* TMS(^{1}H), C_6F_6 (^{19}F) and *para*-dibromotetra-fluorobenzene (^{19}F). Their results can be condensed as given in Table 31

compound	nucleus		temperature region (°C)
TMS	^{1}H	$(+1.44\pm0.33)\cdot 10^{-2} -(0.39\pm0.07)\cdot 10^{-3}t$	−25 to + 25
C_6F_6	^{19}F	$(+2.26\pm0.28)\cdot 10^{-2} -(0.41\pm0.07)\cdot 10^{-3}t$	+ 5 to + 65
PDBTFB	^{19}F	$(-0.92\pm0.37)\cdot 10^{-2} -(0.074\pm0.002)\cdot 10^{-3}t$	+75 to +165

Two comments need to be made at this point. Firstly, the $d\sigma/dt$ data of Table 31 include the intramolecular, the intermolecular and the bulk susceptibility temperature effects. Even after possible correction for the latter, one has still the task of separating the intra- and intermolecular effects. In principle this could be achieved by using the liquids of Table 31 as external references for the measurement of the temperature effect of gases. By doing so at a variety of pressures one could then by extrapolation obtain separately the intra- and intermolecular temperature effects for the gases measured. However, now the second problem emerges, namely that the data of Table 31 carry an imprecision which is too high (by about two orders of magnitude) to make a study as indicated above feasible.

An alternate method for obtaining absolute temperature effects has been described by Meinzer [51]. In his method a reference capillary is placed outside the dewared sample compartment, but still inside the faraday shield. In this way it was hoped that the sample temperature could be varied while the temperature of the reference remained constant. Apart from an uncertainty in this condition there is a major problem, since the magnetic fields, and therefore the resonance conditions for the two tubes, may be rather different (by more than 1 ppm). This by itself is no problem, provided however, that this difference remain absolutely constant during the entire measuring procedure. This latter condition is virtually impossible to guarantee, if only because of the need of reshimming the magnet with varying temperature. After going through all of Meinzer's experimental results we have found that they contain inconsistencies which point to some systematic error in the variable temperature measurements.

We should also like to point to measurements presently under way in the author's laboratory [86] to solve the above noted problems. The method employed is essentially that of Jameson and Jameson, but with two important differences. Firstly a ^{3}He rather than a ^{129}Xe reference resonance is used; since the intermolecular shielding effects of ^{3}He are about a factor of 1000 less than for ^{129}Xe, this will lead to a much more precise extrapolation procedure. Secondly, the intermediate liquid is omitted and the measurements are made directly on gases at variable pressure in a coaxial system.

Finally, there exists in principle yet another method to separate intra- and intermolecular temperature effects, based upon the fact that intermolecular effects are not influenced by isotopic substitution, whereas intramolecular effects are. This method has been used with some success by Mohanty and Bernstein [87] in a temperature and pressure dependence study on the shielding of pure CH_3CF_3 and CD_3CF_3 gases. No similar study on non-polar molecules is known to us (the data of Meinzer [51] on deuterated methanes being considered insufficiently precise).

11.2. $d\sigma_w/dt$ in Gases

The discussion may be started with noting that all macroscopic (dielectric) models for σ_w predict a zero $d\sigma_w/dt$ if the volume is kept constant, as is the case with gas studies. This is a serious shortcoming; it constitutes perhaps the most fundamental criticism of any such model.

The binary collision model on the other hand has an inherent temperature dependence $(d\sigma_w/dt)_v \neq 0$ because of the Boltzmann statistics and of the parameter $y = 2\,(\epsilon/kT)^{1/2}$ which occurs in all such formulations. For example, Eq. (45) leads to a *positive $d\sigma_w/dt$* for all but very small molecules at very high temperatures ($y < 1$). Rummens, Raynes and Bernstein [17] indicated that the inclusion of a repulsion term in the shielding did not noticeably alter the calculated $d\sigma/dt$. In their formulation [Eq. (86)], inclusion of repulsion means an extra factor $[1-\mathscr{H}_{12}(y)/\mathscr{H}_6(y)]$; the ratio $\mathscr{H}_{12}(y)/\mathscr{H}_6(y)$ is always approximately 0.43 and varies little with y. Mohanty and Bernstein [52] studied SF_6, SiF_4 and CF_4 in a variety of non-polar solvent gases at variable temperature. Their experimental results showed $d\sigma_w/dt$ values at least an order of magnitude larger than predicted by Eq. (45). To remedy this discrepancy, they proposed a repulsion factor of the form $[1-C\mathscr{H}_m(y)/\mathscr{H}_6(y)]$ and by data fitting their results they found best values $m = 16$ or 17 and $C = 2.4$; with such values the term $C\mathscr{H}_{16}(y)/\mathscr{H}_6(y)$ is always very close to unity (about 0.95), so that the small temperature dependence of $\mathscr{H}_{16}(y)/\mathscr{H}_6(y)$ is amplified by a factor of about 20.

It would be extremely interesting to test the Mohanty-Bernstein model on $d\sigma_w/dt$ in the liquid state; unfortunately no ^{19}F variable temperature studies on non-polar systems have been reported to our knowledge. For other nuclei the parameter C is likely to be different. Again it is unfortunate that very few variable temperature proton studies in the gas phase have been made; those that have been made are not of sufficient precision to test the Mohanty and Bernstein approach.

Approximate values of $d\sigma_w/dt$, taken from [52] are given in Table 32. These data refer only to the extreme temperatures as indicated in [52].

Table 32. $\Delta\sigma_{lw}/\Delta t$ (ppm cm^3/mole per 100 °C) for some ^{19}F solute gases as measured by Mohanty and Bernstein [52]

Solvent \ Solute	CF$_4$	SiF$_4$	SF$_6$
CF$_4$	+150	+198	+306
SiF$_4$	+122	+340	+330
CH$_4$	+112	+168	+332
Kr	+115	+188	+281
Xe	+232	+360	+306
SF$_6$	–	+280	+288

In spite of the excellent internal agreement, obtained by Mohanty and Bernstein, we feel compelled to repeat (see sections 7.2 and 9.6) an expression of doubt; the model employed leads to values of B which are unrealistically high (by an order of magnitude). There must therefore be another fundamental aspect which has been ignored so far.

In part we believe that higher order collisions may constitute this missing factor. Consider for example the variable temperature data of gaseous ethane [24]. It was found that at all temperatures there is a linear relation between σ_w and ρ, up to $\rho = 0.1$. The temperature dependence of Eq. (45), as applied to ethane can be written in polynomial form as follows:

$$\sigma_{lw} = -18.5 + 0.035\,t - 0.00015\,t^2 \tag{127}$$

Using the above in combination with the experimental data for *liquid* ethane, one finds that all data (liquid + high pressure gas) can be fitted exactly if in addition to Eq. (127) one uses

$$(\sigma_2)_w = +580 + 7t + 0.12\,t^2. \tag{128}$$

It was found that up to $\rho = 0.1$, σ_w is linear with ρ within experimental error (± 0.03 Hz). Also at all gas densities up to $\rho = 0.15$, σ_w was found to vary linearly with temperature. The situation up to this point is therefore completely analogous to that of Mohanty and Bernstein's work. If one now takes $\rho = 0.1$ and $t = 100\,°\text{C}$, Eqs. (127) and (128) give the following results;

$$(\sigma_w)_1 = \sigma_{1w}\rho/M = -0.055 \text{ ppm}$$

$$(\sigma_w)_2 = \sigma_{2w}\rho^2/M^2 = +0.028 \text{ ppm}$$

$$d(\sigma_w)_1/dt = \frac{\rho}{M}\,\frac{d\sigma_{1w}}{dt} = +2 \cdot 10^{-4} \text{ ppm degree}^{-1}$$

$$d(\sigma_w)_2/dt = \frac{\rho^2}{M^2}\,\frac{d\sigma_{2w}}{dt} = +34 \cdot 10^{-4} \text{ ppm degree}^{-1} \tag{129}$$

This example shows that it is quite possible to observe experimentally a strict linear behaviour of σ_w as function of ρ and t, in spite of a considerable contribution of the quadratic $(\sigma_w)_2$ term, while at the same time the temperature dependence of σ_w is even completely dominated by the ternary collisions effect.

Finally the variable temperature work on ^{129}Xe by Kanegsberg *et al.* [80], that on P_4 by Heckmann and Fluck [110] (see also section 14.2) and that on CH_4 by Meinzer [51] should be mentioned.

11.3. $d\sigma_w/dt$ in Liquids

Below, in Table 33 some data as given by Rummens, Raynes and Bernstein [17] are reproduced.

Table 33. $\Delta\sigma_w/\Delta t$ (ppm/100 °C) for some liquids [17]

	temperature range (°C)		$\Delta\sigma_w/\Delta t$ (ppm/100 °C)
$C(CH_3)_4$	− 20	+ 35	+0.016
$Si(CH_3)_4$	− 90	+ 60	+0.021
$Ge(CH_3)_4$	− 90	+ 45	+0.014
$Sn(CH_3)_4$	− 45	+ 55	+0.014
$Pb(CH_3)_4$	+ 5	+100	+0.014
C_6H_{12}	+ 5	+ 80	+0.002
C_5H_{10}	−100	+ 65	$+0.0070 - 73 \cdot 10^{-6}t$

Whereas gases are usually studied at constant volume, liquid studies are normally performed at constant pressure. This therefore causes an extra term $(\sigma_w/\rho)\,d\rho/dt$ in the temperature dependence of σ_w. For neopentane this term is about +0.043 ppm/ 100 °C, i.e. already much larger than the observed value. If, in addition, one invokes the inherent temperature dependence of $\mathscr{H}_6(y)/y^4$ of Eq. (45), this adds another positive term equal (for neopentane) to +0.068 ppm/100 °C. The sum of these two terms is of the correct sign, but an order of magnitude too large. Rummens *et al.*

[*17*] tried to remedy this point by making their universal fudge factor K_6^g temperature dependent. They quote a "best" result of $K_6^g = 1.46 + 0.0064t$ but even with this the calculated $d\sigma_w/dt$ values are still in poor agreement with experiments.

Following the ideas exposed in the previous section one might try to incorporate ternary collisions. However, for all liquids *except* ethane, the σ_w (liquid) is larger than expected from extrapolation from a low pressure σ_w versus ρ plot. Furthermore one now needs a *negative* temperature dependence of σ_2, contrary to what was found for ethane. Rummens [*24*] has argued already that the above is not necessarily contradictory. He starts by noting a fundamental equivalence between σ_1 and the second virial coefficient $B(T)$ for the equation of state and proceeds to assume that a similar equivalence is likely to exist between σ_2 and $C(T)$, the third virial coefficient. He then shows that for ethane the reduced temperature $T^* = kT/\epsilon$ is larger than unity (in the NMR experimental range) i.e. a range where $B^*(T^*)$ is negative with a positive temperature coefficient and $C^*(T^*)$ is positive, with a positive quadratic temperature coefficient exactly like the parameters σ_1 and σ_2. For molecules like those of Table 33, however, T^* is usually much less than unity, where both $C^*(T^*)$ and $B^*(T^*)$, and therefore also both σ_{2w} and σ_{1w}, are negative, thus *adding* to each other. In addition, it appeared that the ratio $C^*(T^*)/B^*(T^*)$ is nearly constant in this T^* region, hence providing a welcome "explanation" for the apparent universality of K_6^g.

The above ideas have not been followed up, but particularly after the appearance of the Mohanty and Bernstein study [*52*] such an exploration would seem eminently useful.

Spanier and Malinowski [*88*] have measured $d\sigma_w/dt$ for the ^{19}F and ^{1}H shifts of *para*-difluorobenzene. They did these measurements relative to the ^{1}H signal of gaseous C_2H_6, assuming that the temperature effect of the latter is zero. While this is not exactly correct, the relative magnitude of this error is small in the ^{19}F case, because σ_w for ^{19}F (and hence its temperature dependence) is so much larger than for ^{1}H. Their results are given in Table 34.

Table 34. ^{19}F and ^{1}H chemical shift (ppm) of *para*-difluorobenzene as function of temperature relative to that at $0\,^\circ$C [*88*] (positive sign indicates up-field shift)

t (°C)	^{19}F (ppm)	^{1}H (ppm)
−5	+0.001	+0.021
0	0.000	0.000
10	+0.013	−0.032
25	+0.046	−0.066
40	+0.087	−0.091
55	+0.132	−0.112
70	+0.180	−0.131
85	+0.235	−0.143
100	+0.302	−0.142

It may be noted that the temperature dependence is not linear in this case. The data can be condensed as $\sigma(t) = \sigma(0) + 1.63 \cdot 10^{-3} t + 1.41 \cdot 10^{-5} t^2$ for the ^{19}F shifts and $\sigma(t) = \sigma(0) - 2.84 \cdot 10^{-3} t + 1.41 \cdot 10^{-5} t^2$ for the ^{1}H shifts. The data were corrected for bulk susceptibility at each temperature, using the spinning side band technique of Spanier, Vladimiroff and Malinowski [89] (see section 15.3). This is the main reason why the results for $d\sigma_w/dt$ appear an order of magnitude smaller than those of Table 31. Surprising is the negative sign for $d\sigma_w/dt$ for ^{1}H and the large magnitude of the latter. We feel that there must be systematic error here, either caused by the σ_b correction, or by the assumption that the 8 atm of ethane, used as a reference, would have a zero $d\sigma_w/dt$.

Finally, the study on $d\sigma_w/dt$ of liquid white phosphorus by Heckmann and Fluck [110–112] should be mentioned; this will be discussed in section 14.2.

Chapter 12. Factor Analysis

12.1. Introduction to the Formalism

Factor analysis, as introduced by Malinowski and co-workers [90, 91, 92, 93], aims to provide a means of dividing the total experimental medium shift into its contributants σ_w, σ_a, σ_E etc. As such, it is of little direct bearing on the subject of this review, which restricts itself to systems where there is only one contribuant; σ_w. At the same time it is clear that our preoccupation with σ_w has as one of its goals the development of a model for σ_w which allows us to *calculate* σ_w with good confidence in those systems where the medium shift is composed of two or more terms. Factor analysis therefore seems to provide an alternative way of separating these effects.

The technique, although described in detail by Weiner, Malinowski and Levinstone [91] will be recounted below.

Assume a matrix of medium shifts S of m solutes in n solvents. It is now assumed that each such shift (element of the S matrix) can be written as

$$S_{ik} = \sum_{j=1}^{r} u_{ij} v_{jk} \tag{130}$$

The u_{ij} ($i = 1, 2, 3, \ldots, m$) are solute factors, i.e. numbers exclusively determined by parameters of the solute. Similarly v_{jk} ($k = 1, 2, \ldots, n$) are pure solvent factors. The summation in Eq. (130) is over $j = 1, 2, \ldots, r$ where in general $r = m$ or $r = n$ *(vide infra)*. The shift matrix [S] of order $m \times n$ is next symmetrised by multiplication by its transpose $[S]'$. If this is a pre-multiplication, a correlation matrix [C] of order $n \times n$ arises.

$$[C] = [S]'[S] \tag{131}$$

In this case $r = n$. It is also possible to post-multiply so that a $m \times m$ symmetrical matrix arises with $r = m$. The matrix $[C]$ of $r \times r$ can now be diagonalised by finding a matrix $[B]$ such that

$$[B]^{-1}[C][B] = [\lambda_j \delta_{ij}] \tag{132}$$

where δ_{ij} is the Kronecker symbol and λ_j $(j = 1, 2, \ldots, r)$ are the eigenvalues. With each λ_j there is associated an eigenvector $\{B_j\}$, i.e. the j-th column of $[B]$, so that

$$[C]\{B_j\} = \lambda_j\{B_j\} \tag{133}$$

The vectors $\{B_j\}$ are orthogonal (so that $[B]^{-1} = [B]'$) and they can therefore be used as a basis for $[C]$. One may now write

$$[B]^{-1}[C][B] = [B]^{-1}[S]'[S][B] = [B]'[S]'[S][B] = [\lambda_j \delta_{ij}] \tag{134}$$

It can be seen that Eq. (134) can be satisfied by introducing a matrix $[U]$ such that

$$[U] = [S][B] \rightarrow [U]' = [B]'[S]' \tag{135}$$

The condition $[S] = [U][V]$ is therefore fulfilled if additionally we have

$$[V] = [B]' \tag{136}$$

Provided one is able to solve the eigenequation of order λ^r, one finds r eigenvectors which together constitute $[B]$ and $[B]'$. Each element in $[S]$ is then the sum of r products of the form $u_{ij}v_{jk}$ as required by Eq. (130). So far this is purely mathematics; if one had started with an $r \times r$ $[S]$ matrix, the result would be an exact duplication of that matrix [assuming Eq. (130) holds true].

In attempting to make a physical correlation one then orders the λ_j according to magnitude and asks how many λ_j's are needed to reproduce $[S]$ *within experimental error*. We pause here to make three remarks. Firstly it is not always easy to say what constitutes the experimental error limit. For example, in gas-to-liquid shifts the correction for bulk susceptibility constitutes a major uncertainty (of perhaps up to 0.02 ppm.); this, however, is a systematic error for each and any solvent and it simply becomes absorbed in the solvent factor for that solvent. The same fate will befall any other error which is systematic for any solute or solvent. It is therefore the repeatability or precision as determined by *random* errors that has to be considered. Secondly, if r factors are sufficient to reproduce the matrix within (random) error, this only means that there are *at least r* physical phenomena at work; there may be more of even substantial magnitude which may go unnoticed. This may come about if such additional effect or term is nearly proportional, in its *numerical values of the solvent matrix under study* to one of the other terms. If, by writing such an extra effect in terms of a "best" constant times one of the other terms, and if by so doing the errors introduced are of the same order of magnitude as the experimental error, such extra term will not be found by factor analysis. Thirdly, one has to have a certainty that all contributing terms can be written in product form. If this is not the case, and if for example each term such as σ_w, *in itself* requires the sum of two or more products, then of course factor analysis fails to give any physical insight whatsoever.

To continue the description of the Malinowski technique, a rotation of the U and V matrices is required to make them identifiable with physically significant factors. This rotation is carried out over the *reduced* matrices $[U]_{rp}$ and $[V]_{pr}$ of order $r \times p$ and $p \times r$ resp., where $p \leqslant r$ is the minimum number of factors required. If the rotation matrix is $[R]$, and $[U]$ and $[V]$ are the rotated reduced matrices, one has

$$[\bar{U}] = [U][R] \text{ and } [\bar{V}] = [R]^{-1}[V]. \tag{137}$$

In order to know how to rotate it is necessary to *guess* either the "true" U matrix $[\bar{\bar{U}}]$ or the "true" solvent matrix $[\bar{\bar{V}}]$. This can only be done by the injection of assumedly correct models for each contributing effect and is only possible if these models allow the writing of each term as a product of solute parameters times solvent parameters. Rotating $[U]$ in such a manner that the differences $\bar{U}_{ik} - \bar{\bar{U}}_{ik}$ are minimised according to a least square criterion then leads to [91]

$$\{R\}_j = [(1/\lambda_j)\delta_{jk}][U]' \{\bar{\bar{U}}\}_j \tag{138}$$

for each column of the rotation matrix $[R]$. The sought-after matrix $[\bar{U}]$ then follows from

$$\{\bar{U}\}_j = [U]\{R\}_j \text{ and } [\bar{V}] = [\bar{U}]^{-1}[S] \tag{139}$$

It is equally possible to rotate the solvent matrix according to

$$\{R\}'_j = [(1/\lambda_j)\delta_{ij}][U]' \{\bar{\bar{U}}\}'_j \tag{140}$$

Weiner, Malinowski and Levinstone [91] adopted the latter technique for an S matrix where the solutes and the solvents were halo-methanes (including CH_4 and CH_3CN and also CS_2 as a solvent). They found three factors to be "sufficient", using a restricted submatrix of S. We note, however, that the deviations (experimental minus predicted) averaged about 1 Hz, with a maximum of about 2 Hz, and about 50% of the data having a deviation larger than 1 Hz. Since the data in question were internally referenced with accurate calibration of the frequencies, we believe that the experimental error could have been no larger than ±0.2 Hz. In our opinion, therefore, at least *four* factors would have been required. This shortcoming turns up in amplified form, when the authors perform a "best" rotation into three solvent factors, for which they take the shifts (for each solute) in CH_3CN, CCl_4 and CH_2Br_2. Using this basis, shifts, which were not used in the determination of the solvent factors, were predicted and compared to the experimental data. The authors state that "the agreement, although slightly beyond experimental error, is reasonably satisfactory". We noted, however, that the deviations now range all the way up to 7.0 Hz (0.12 ppm). The standard deviation is ±2.4 Hz (±0.04 ppm) *i.e.* an order of magnitude larger than the experimental error. In our opinion the rotation into solvent shifts, at least with the basis employed, should have been termed "unsuccessful". The authors did find that the gas phase shift is a good factor, which is as it should be. But this leaves only two factors in a system where at least three other factors (σ_w, σ_a and σ_E) are at work. It might therefore well be, that inclusion of one extra factor could produce a much more satisfactory result.

12.2. Medium Effects as Product Functions

One very basic requirement for the successful application of factor analysis to solvent effects is that each contributing term (σ_w, σ_a, σ_E etc.) must have the form of a solute factor times a solvent factor. That such is the case has been indicated by several authors [7, 43, 90, 93]. Malinowski and Weiner [90] write for example Eq. (76) in the following form:

$$\sigma_w = K_6^g \frac{-\pi NB\alpha_2 I_2 \mathcal{H}_6(y)}{V_2 r_0^3 y^4} S_6^g$$

$$= \left[\frac{-B}{r_{01}^{3/2}} \left(\frac{\mathcal{H}_6(y_1)}{y_1^4}\right)^{1/2} S_6^g \right] \cdot \left[\frac{\pi K_6^g N\alpha_2 I_2}{V_2 r_{02}^{3/2}} \left(\frac{\mathcal{H}_6(y_2)}{y_2^4}\right)^{1/2} \right] \tag{141}$$

where the following assumptions have been made:

$$y = (y_1 y_2)^{1/2} \tag{142}$$

$$r_0 = (r_{01} r_{02})^{1/2} \tag{143}$$

$$\mathcal{H}_6(y) = \mathcal{H}_6(y_1)^{1/2} \cdot \mathcal{H}_6(y_2)^{1/2} \tag{144}$$

We will later comment on the above assumptions but we note already, however, that S_6^g is not a pure solute parameter, since one has $q_0 = d/r_0 = 2d/(r_{01} + r_{02})$ [see also Eqs. (78) and (48)].

For non-polar solutes in polar solvents there is a term σ_{E2} wich can be written as [93]:

$$\sigma_{E2} = \sigma_w \cdot \frac{2\mu_2^2}{3\alpha_2 I_2} \tag{145}$$

where the site factor S_{E2} (see ref. [94] and the scale factor have been taken equal to those of the σ_w effect. Therefore the σ_{E2} term can never be identified by factor analysis; if present, one can at best find the solvent factors corresponding to ($\sigma_w + \sigma_{E2}$), while the solute factors of σ_w and σ_{E2} are identical.

The σ_a term presents a slight problem because in the binary gas model σ_a should be equal to zero (except when both the solute and solvent are polar). Nevertheless in the liquid σ_a may be appreciable, as has been found experimentally. Various early model calculations by Homer [95], Schug [96] and Abraham [97] suggested that σ_a

should be solute independent, so that for any system where the solutes are non-polar, one may write

$$(s_{ij})_{g \to \varrho} = (u_i)_w (v_j)_w + 1.(v_j)_a \tag{146}$$

Therefore, medium shifts of any solute in a number of solvents, plotted against the medium shifts of another chosen solute in the same solvents should result in a straight line for each of the solutes.

12.3. Applications of Factor Analysis

Malinowski and Weiner [90] used basically the above technique, although they used externally referenced experimental data for solutions, so that the gas shift is another complicating factor. In such plots the slope is given by

$$m = \frac{\delta_{ij} - \delta_{ik}}{\delta_{aj} - \delta_{ak}} = \frac{(\sigma_{ij})_w - \sigma_{ik})_w}{(\sigma_{aj})_w - (\sigma_{ak})_w} \tag{147}$$

where j, k are solvents, i is the solute to which the line pertains and a is the "reference solute" of the plot. If indeed each σ_w term can be written as a product $(\sigma_{ij})_w = (u_i)_w \cdot (v_j)_w$ one obtains

$$m = (u_i)_w / (u_a)_w \tag{148}$$

which is a ratio of solute factors only, and a pure constant, independent of the solvents chosen.

The above example is interesting because of several reasons;

(i) The graphs so determined are indeed very nicely linear except for points relating to anisotropic solvents.

(ii) The slope as calculated per Eq. (148), taking the solute factors of Eq. (141), is close to the experimental slope (in spite of the fact that the authors apparently set $q_0 = d/r_0 = 2\, d/(r_{01} + r_{02}) \approx 2\, d/2\, (r_{01})^{1/2}$ for all solute-solvent systems).

(iii) Given at least one gas phase shift, the gas phase shifts of the other solutes can be determined from the experimental δ_{ij} versus δ_{aj} plots.

(iv) It is possible to determine the σ_a contribution provided that $\sigma_a = 1.(\sigma_j)_a$ and provided the points for anisotropic solvents do not happen to fall on the line. The technique is to subtract a constant amount (i.e. $= \sigma_a$) from both the δ_{ij} and the δ_{aj} coordinate such that the point so obtained falls on the line. For the CH_3 and the CH_2 signals of 1,1,3,3-tetramethylcyclobutane as well as for cyclohexane this technique gave σ_a's for benzene of +0.23, +0.42 and +0.43 ppm respectively for an average of +0.36 ppm. In a similar fashion they found $\sigma_a(CS_2) = -0.015$ ppm and σ_a (CH_3CN) = -0.005 ppm with all data ± 0.005 ppm precise.

(v) From the same plots solvent factors can also be obtained. If d_j, d_k are the distances along any line from the gas point to the respective δ_{ij}, δ_{ak} and δ_{aj}, δ_{ak} points, then, assuming a product rule for σ_w one has

$$\frac{d_j}{d_k} = \frac{(v_j)_w}{(v_k)_w} \tag{149}$$

The authors quote a good agreement between experimental and predicted [on the basis of Eq. (141)] ratios for C_5H_{10}/CCl_4 and C_6H_{12}/CCl_4. No comments are given for the other solvents. One might speculate that for these other solvents (which were all polar) the agreement was not so good. It would have been interesting to see whether the inclusion of the σ_{E2} term of Eq. (145) would have produced better results.

Weiner and Malinowski [92] also determined σ_a effects using factor analysis on non-polar solutes in a number of solvents of various nature. They assume that the externally referenced shifts (relative to R) of solute i can be written as follows:

$$\delta_{ij}^R = \delta_{i\,gas}^R \cdot 1 + (u_i)_w(v_j)_w + 1 \cdot (v_j)_a, \tag{150}$$

where we note that the Van der Waals term includes the σ_{E2} effects. The above is also true for any chosen reference solute for which the authors took CH_4

$$\delta_{ij} = \delta_{ig}^R \cdot 1 + \frac{(u_i)_w}{(u_{CH_4})_w} \cdot \delta_{CH_4,j}^{CH_4} + \left(1 - \frac{(u_i)_w}{(u_{CH_4})_w}\right)(v_j)_a \tag{151}$$

The suspected solvent factors are therefore respectively unity (with the gas phase shifts of solutes i as coefficients), the CH_4 gas-to-solution shifts $\delta_{CH_4,j}^{CH_4}$ and the solvent anisotropy $(v_j)_a$. Note that the sum of the two coefficients of the latter two vectors is unity. The first two vectors are known, but the third presented a problem since a rather wide range of estimated σ_a effects exists in the literature, even for C_6H_6 and CS_2. The authors therefore took $(v_{CCl_4})_a = 0$ and used variable parameters for $(v_{C_6H_6})_a$ and $(v_{CS_2})_a$, then rotated into these vectors and tried to find the best rotation that would also fulfil the unity sum rule for the last two coefficients of Eq. (151). They found that this was achieved with $\sigma_a(C_6H_6) = +0.40$ ppm and $\sigma_a(CS_2) = -0.015$ ppm (as compared to +0.36 and −0.015 ppm obtained with the earlier mentioned graphical method). The excellent agreement is somewhat misleading, however. In both studies the CH_4 shifts are used as basis, while the solutes and solvents of the two studies are rather similar molecules and partly the same.

The method of Bernstein [93] is similar to the graphical methods of Malinowski *et al.*, but there are characteristic differences. Bernstein starts by pointing out that in gases not only is $\sigma_a = 0$ but that in many cases the sum of three other terms σ_w, σ_{E2} and σ_E can be combined into one single product $u_{ik}v_{kj}$. On the basis of a solute number for ^{19}F of CF_4 equal to 1, he proceeds to find other solute numbers by plotting gas phase shifts (actually second virial coefficients σ_1) for these solutes in a number of solvents against the CF_4 shifts in the same solvent gases. These plots go through the origin and have as slope the solute number of the solute in question. Having these ^{19}F solute numbers, it is a simple matter to find "best" solvent numbers which are then in turn used to find the solute number for ^{1}H of CH_4 in these same solvents and hence all other ^{1}H solute numbers. Bernstein then proceeds to take these gas phase numbers to the liquid phase; the solute numbers without change, but the solvent numbers multiplied by K/V_2 where K is a scale factor such as the

K_8^g of Eq. (141). This carry-over is practicable because some solutes (CH_4, $H_2C{=}CF_2$ and HCl) were used both in the gas phase and in the liquid phase studies, and from here on an extended set of solute and solvent numbers for liquids is derived. For systems having anisotropic solvents a new term σ_a has to be added, which Bernstein again assumes to be independent of solute. He then shows that a plot of gas-to-liquid shifts of solutes i in a solvent j, plotted against the solute number of i should give a straight line with the solvent number of j as slope and σ_a as intersept. The following results were obtained; $\sigma_a(C_6H_6) = +0.625$ ppm, $\sigma_a(CS_2) = -0.05$ ppm, $\sigma_a(CN_3CN) = -0.05$ ppm, $\sigma_a(CHCl_3) = 0$, $\sigma_a(\text{acetone}) = +0.13$ ppm, $\sigma_a(C_6H_5NO_2) = +1.00$ ppm. Bernstein's work is remarkable because he treats σ_w, σ_{E2} *and* σ_E as linearly *dependent* factors. Since he finds internally consistent results, this throws some doubts on our earlier remark *re* the factor analysis work by Weiner, Malinowski and Levinstone [*91*] (see section 12.1); perhaps after all, these authors used the correct number of factors, which then would have been the gas-phase shift, the neighbour anisotropy and the combined $\sigma_w + \sigma_{E2} + \sigma_E$ effect. Note, however, that Bernstein's σ_a values are rather different from those obtained by Malinowski *et al* [*91, 92*]. Yet each of these studies by themselves looks impressive in its selfconsistency and in the apparent precision of the various plots or mathematical rotations.

Factor analysis on ^{19}F medium shifts has been studied by Abraham *et al.* [*98, 99*] and on ^{13}C and ^{29}Si medium effects by Bacon and Maciel [*100*]. These studies will be discussed in Chapter 13 and section 14.1 respectively.

12.4. Criticism of the Factor Analysis Method

The present author has serious objections against the use of factor analysis in NMR solvent effects. That this idea is not entirely preconceived might follow from the following argumentation.

(i) The assumption that $u_i = 1$ for the σ_a effect is probably not a good approach. Rummens [*34*] has already shown that even for spherical molecules σ_a is proportional to $V_1^{-1/6}$. More seriously, there appears to be a rather strong site effect for σ_a if the solute is non-spherical. Rummens and Louman [*58*], using non-polar solutes in non-polar solvents, estimate σ_a in C_6H_6 to vary from $\sigma_a = +0.34$ ppm for $CH_3CH{=}C{=}CHCH_3$ in C_6H_6 to $\sigma_a = +0.56$ ppm for CH_3 in the same molecule. Admittedly the basis for these predictions is a calculation of σ_w according to a somewhat dubious model, but nevertheless the differences found for σ_a seem to be much too large to be attributable to model errors.

(ii) Louman's factor analysis study on his complete set of non-polar, non-polar gas-to-liquid shifts [*57*] is very revealing. Using the matrix containing isotropic solvents only (see Tables 16 and 17), *one* factor appears to be sufficient but there are some rather large discrepancies (up to 0.04 ppm). This matrix provided a set of Van der Waals solute and solvent numbers. Plotting the shifts in CS_2 versus the solute numbers resulted in an excellent straight line from which $\sigma_a = -0.126$ ppm was derived. It is therefore apparent that the anisotropy contributions derived depend rather strongly on the basis set of data employed, which is contrary to the basic assumption that $(u_i)_a = 1$. Louman tried similar plots for C_6H_6, $p\text{-}C_6H_4(CH_3)_2$ and mesitylene. These plots were utterly non-linear as is apparent from Figure 8. This is positive proof of the non-validity of at least one of the assumptions made (either $(u_i)_a \neq 1$ or $s_{ij} \neq \Sigma u_{ik}v_{kj}$). Note that if only the six points for CH_2 and CH_3 of the tetra ethyl compounds were used, a decent straight line with $\sigma_a(C_6H_6) = +0.51$ ppm results, the latter in between the results given by Bernstein (+0.65) and by Malinowski *et al.* (+0.42). If other solute points of Figure 8 are selected, other, rather

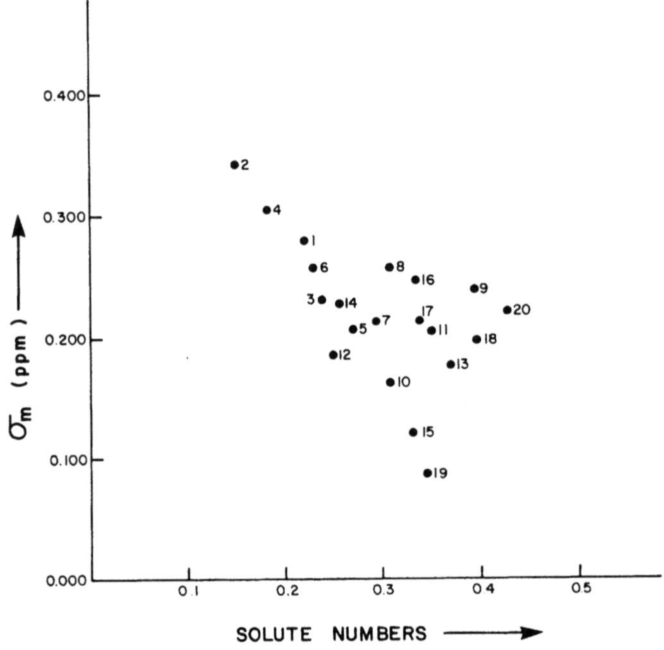

Fig. 8. Graph of the medium shifts in benzene *vs* solute numbers, the latter being obtained from the data of Tables 26 and 27. Data according to Louman [57] and Rummens and Louman [58]

different σ_a values result. Louman tried also a full factor analysis treatment on his non-polar, non-polar anisotropic matrix. He found the following eigenvalues $\lambda_1 = 2.449$, $\lambda_2 = 0.0122$ and $\lambda_3 = 0.0017$. This means that only *one* factor is predominant. Indeed, using λ_1 only, one can nearly reproduce the experimental matrix with a largest deviation of 0.04 ppm, i.e. just as good as the non-polar, non-polar isotropic matrix with one factor. In other words, if there are two factors in the non-polar, non-polar anisotropic matrix, then there are also two factors in the non-polar, non-polar *isotropic* matrix. *Vice versa,* if one insists that σ_w can be expressed in one factor, then also $\sigma_w + \sigma_a$ needs only one factor. It seems therefore clearly indicated that σ_w and σ_a are strongly dependent and can never be separated by factor analysis. Louman also tried to rotate the solute factors by using for the $\{\bar{\bar{U}}_1\}$ vector the solute numbers from his study with isotropic solvents, and for $\{\bar{U}_2\}$ either unity, or $V_1^{-1/6}$ or a number of other possibilities. He found that such a rotation, if made, resulted in a rotated vector $\{\bar{\bar{U}}_1\}$ always very different from $\{\bar{U}_1\}$, which is completely contrary to the basic assumptions. One might argue that perhaps the largest eigenvalue corresponds to the σ_a effect and not to the σ_w effect, but since the original $\{U_1\}$

vector has numbers that vary rather widely (by more than a factor 3) this is again unlikely, unless one assumes that the solute vector for σ_a can have elements with that kind of variability. If one insists in using solute numbers $\{U_1\}$ obtained from the study with isotropic solvents and a solute vector $\{U_2\} = 1$, the agreement between experiment and prediction becomes very poor as shown again by Louman (standard deviation ± 0.07 ppm, largest deviation 0.12 ppm). Clearly something is very wrong.

(iii) It may be argued that the assumption $\sigma_w = (u_i)_w \cdot (v_j)_w$ is wrong. Consider for example the way this has been indicated in terms of the basic binary collision model as per Eqs. (141–144). In particular take Eq. (143). Clearly,

$$\left(\frac{r_1 + r_2}{2}\right)^{-3} = r_1^{-3/2} \cdot r_2^{-3/2}$$

is only a first approximation; it is in error by about 50% if $r_1/r_2 = 3$ (which corresponds roughly to $SiEt_4$ in CS_2!). A much better approximation would be (with $r_1 > r_2$)

$$\left(\frac{r_1 + r_2}{2}\right)^{-3} = r_1^{-3/2} \cdot r_2^{-3/2} - 0.046\, r_1^{1/3} \cdot r_2^{-10/3} \tag{152}$$

which is correct to within 5% in the range $r_1/r_2 = 1$ to $r_1/r_2 = 3.5$. Note that the second term is about 33% of the first term for $r_1/r_2 = 3$, which is certainly not negligible. Consider next the site factor S_6^g, which according to Malinowski and Weiner [90] is part of the solute factor. The site factor contains the solute parameter d as well as the Lennard-Jones r_0 parameter, the latter being both a solute and a solvent parameter ($2\, r_0 = r_{01} + r_{02}$). However, one might write

$$S_6^g (d, r_0) \approx [S_6^g (d, r_{01}) \cdot S_6^g (d, r_{02})]^{1/2} . \tag{153}$$

It can be shown that the above approximation is equivalent to the approximation of Eq. (143). Note that $S_6^g(d, r_{01})$ is now a pure solute term, but that $S_6^g(d, r_{02})$ is still a function of solute and solvent. If one now considers the first two terms of $S_6^g(d, r_{02})$, one may write

$$S_6^g(d, r_{02}) \approx 1 + C \frac{d^2}{r_{02}^2} \approx \frac{1}{r_{02}^2} (r_{02}^2 + C \cdot d^2) \tag{154}$$

which, if r_{02}^2 and $C \cdot d^2$ are not too different, gives

$$\frac{1}{r_{02}^2} \cdot 2 \sqrt{r_{02}^2 C d^2} = \frac{2\, d}{r_{02}} \sqrt{C} \tag{155}$$

which is in the required product form. Each further term in $S_6^g(r_{02})$ may be treated in a similar fashion. The result is always a pure product in d and r_{02}, but one has to use a geometric mean for an algebraic mean as many times as there are important terms in the series development for S_6^g. Again it is found that S_6^g is not a pure product, but that one may write it as the *sum* of two or more (depending on the accuracy required) pure products.

Chapter 13. ^{19}F σ_w Studies

The first measurement of a σ_w effect in ^{19}F NMR was reported in 1958 by Evans [14]. For CF_4 he reported a gas-to-infinite-dilution-in-CCl_4 shift of 9.22 ppm (7.76 ppm if corrected for solvent susceptibility). Almost at the same time Glick and Ehrenson [101] reported 9.25 ppm for the same experiment. In a subsequent paper [102], Evans gave also the following additional σ_w data; CF_4 in C_7H_{16} 6.06 ppm and SiF_4 in CCl_4 11.34 ppm. For gaseous CF_4 he found $\sigma_w = (1.15 \pm 0.15) \cdot 10^{-2}$ ppm/atm at 21 °C, which corresponds to $\sigma_1 \approx 255$ ppm mole^{-1} cm^3. Evans pointed to the Van der Waals forces as the probable source for these medium shifts. Glick and Ehrenson [101] also pointed out that these shifts were much too large to be described to bulk susceptibility, and noted a parallel between these solvent shifts (mostly of the polar-polar type) and solvent polarisability. Much of the gas phase ^{19}F work by Bernstein and his co-workers has alsready been discussed in sections on the repulsion effect (section 7.2), the B parameter (section 9.6) and the temperature dependence of σ_w (section 11.2). In the present chapter we will therefore only tabulate the experimental data.

Table 35. Observed $-\sigma_{1w}$(ppm cm^{3}mole^{-1}) data of Petrakis and Bernstein [42] at 30 °C

solute	solvent				
	CF_4	CH_4	SiF_4	SF_6	C_2H_6
CF_4	91	136	96	196	–
SF_6	–	181	110	206	–
SiF_4	–	–	209	271	342

The large discrepancy between the experimental results of Tables 35 and 36 is particularly noteworthy. Mohanty and Bernstein ascribe it to the superior gas-handling techniques of their study, but even so, this constitutes a warning sign that gas phase data must be considered with suspicion and should not be taken for granted until and unless reproduced by other workers or at other times with different techniques.

Table 36. Observed $-\sigma_{1w}$(ppm cm^3 mole^{-1}) data of Mohanty and Bernstein [52] at 30 °C (for $\Delta\sigma_{1w}/\Delta t$ see Table 32)

solute	solvent					
	CF$_4$	CH$_4$	SiF$_4$	SF$_6$	Kr	Xe
CF$_4$	198	222	239	316	247	458
SF$_6$	239	284	284	358	291	489
SiF$_4$	257	320	355	402	347	621

Raynes and Raza [103] have reported a number of ^{19}F σ_w shifts. Their results are given in Table 37.

Table 37. ^{19}F $-\sigma_w$(ppm) data at 35 °C according to Raynes and Raza [103]

solute	solvent				
	SiEt$_4$	SnEt$_4$	Sn(CH$_3$)$_4$	CCl$_4$	SiCl$_4$
CF$_4$	6.00	6.26	6.82	7.60	6.85
SF$_6$	6.35	6.70	7.05	7.89	7.03
SiF$_4$	14.37	15.97	10.11	11.15	10.10
p-C$_6$H$_4$F$_2$	7.21	7.43	7.74	8.27	7.96
C$_6$F$_6$	7.16	7.21	7.89	8.81	8.45

Raynes and Raza noted that for the solutes CF$_4$, SF$_6$ and SiF$_4$ there is a striking parallel; the solvent shifts of SF$_6$ and SiF$_4$ can be obtained quite accurately by multiplying those of CF$_4$ by a factor 1.06 and 1.50 respectively. This means, incidentally, that for that matrix the solvents effects can be written as a product function (see Chapter 12). In fact the same is true, be it to somewhat lesser extent, for the solutes p-C$_6$H$_4$F$_2$ (factor 1.15 ± 0.05) and C$_6$F$_6$ (factor 1.18 ± 0.03). The authors noted also that the ^{19}F shifts of Table 37 are roughly proportional to the ^{1}H shifts of the corresponding proton compound (CH$_4$ vs CF$_4$, C$_6$H$_4$F$_2$ vs C$_6$H$_4$F$_2$) of Table 15. They see this as evidence of the same physical phenomenon at work in both cases. This is somewhat in variance with their conclusion that with ^{19}F it is primarily the paramagnetic part which contributes to the Van der Waals effect, whereas with ^{1}H the diamagnetic contribution is more important. It may be noted, however, that the solvent CCl$_4$ appears to form an exception in the above noted relation.

Abraham, Wileman and Bedford [98] more recently embarked on a comprehensive study of ^{19}F medium effects. From their work we again selected only those gas-to-liquid shifts which are purely Van der Waals. These results are reproduced in Table 38.

Table 38. Gas-to-liquid shifts $-\sigma_w$(ppm) of ^{19}F at 39–42 °C. (0.1 to 2 atm and 1–2 mole %), according to Abraham *et al.* [*98*]

solute \ solvent	CCl$_4$	n-C$_7$H$_{16}$	C$_6$H$_{12}$	n-C$_6$H$_{14}$	n-C$_5$H$_{12}$	n-C$_6$F$_{14}$	C$_4$F$_8$
CF$_4$	7.60	6.08	5.94	5.77	5.44	3.34	3.04
C$_4$F$_8$	6.94	5.62	5.49	5.34	4.96	3.08	2.87
n-C$_6$F$_{14}$ | $-$CF$_3$	5.58	4.48	4.44	4.28	4.06	2.53	2.32
n-C$_6$F$_{14}$ | α–CF$_2$	4.33	3.53	3.51	3.37	3.19	2.00	1.89
CF$_3$C≡CCF$_3$	6.51	5.28	5.18	5.00	4.73	2.72	2.50
trans-CFCl=CFCl	6.91	5.83	5.78	5.54	5.25	2.83	2.65
C$_6$F$_6$	9.21	7.72	7.62	7.38	6.90	3.87	3.72

The authors discuss the results mainly in terms of factor analysis; this aspect has already been treated in Chapter 12. The results of the factor analysis are of about equal quality or slightly better as found with ^{1}H shifts; deviations of up to 0.3 ppm (i.e. 5 to 8% of the total σ_w shift) are found. The partial improvement may be due to the fact that ^{19}F σ_w shifts are large so that experimental precision plays an almost negligible role. Abraham *et al.* are more optimistic in their interpretation; according to them the deviations of 0.2 and 0.3 ppm are due to assumed specific interactions between fluoroalkanes and hydrocarbons. In addition, the authors attempted to correlate the solvent numbers, obtained from the factor analysis, with theoretical expressions such as those of Howard, Linder and Emerson [Eq. (19)] or a proportionality with $\alpha_2 I_2^{1/2} V_2^{-1}$ (see top line of Table 7). In both cases they find a roughly linear correlation, which straight lines did, however, not go through the origin (*i.e.* the gas-point). Unfortunately, the authors did not attempt to use the binary collision model. This is the more remarkable since they did find that the solvent numbers, plotted versus $n_2^2 - 1$, produces a straight line *through the gas point* ($n_2^2 - 1$ is always roughly proportional to α_2/V_2).

In Table 39 some early data of Frei and Bernstein [*104*] are given.

Table 39. $-\sigma_w$ data (ppm) for ^{19}F of CF$_4$, according to Frei and Bernstein [*104*]

solvent	$-\sigma_w$(ppm)
CCl$_4$	7.80
n-C$_6$H$_{14}$	5.77
n-C$_7$H$_{16}$	6.08
C$_6$H$_6$	6.61
CS$_2$	7.90

These data correspond well with those of Tables 37 and 38. Cyr, Raza and Reeves [*105*] have recently studied C$_6$F$_6$ in a variety of solvents. From their work we quote

C_6F_6 in CCl_4 $\sigma_w = -9.09$ and C_6F_6 in cyclohexane $\sigma_w = -7.40$ ppm, in reasonable agreement with the data of Abraham *et al.* (Table 38). The earlier ^{19}F work by Emsley and Phillips [106] we find less reliable. Not only are all their data relative to an internal standard, but in their study of non-polar solutes they mixed all the solutes together in each solvent (2 mole % each of C_6F_6, C_6F_{12} and $C_6F_4Cl_2$ and 5 mole % $CFCl_3$). Not only is now the conversion to external standard virtually impossible, but even their internal data are not trustworthy. For example they quote 163.08 ppm for the internal difference between $CFCl_3$ and C_6F_6 in CCl_4, while Abraham *et al.* [98] measured 162.37 ppm for the same property.

In Table 40 we have collected the available data on ^{19}F gas phase shifts of non-polar compounds.

Table 40. Gas phase shifts $-\sigma_0^*$ of ^{19}F, relative to CF_4

compound	$-\sigma_0^*$ (ppm)	footnote
SiF_4	-104.29	a
	-104.19	b
	-104.18	c
C_6F_6	-101.90	d
	-101.66	b
C_4F_8	-72.03	d
n-C_6F_{14} (α-CF_2)	-61.56	d
p-$C_6H_4F_2$	-58.24	b
$CFCl=CFCl$	-56.66	d
n-C_6F_{14} (CF_3)	-17.56	d
CF_4	0.000	–
$CF_3C\equiv CCF_3$	$+9.46$	d
SF_6	$+119.44$	a
	$+119.451$	c
	$+119.50$	b
	$+119.51$	e
F_2	$+491.8$	c

[a]) At $t = 30\,°C$ relative to liquid ethyltrifluoroacetate (uncorrected) extrapolated to zero pressure. Data from Petrakis and Bernstein [42]. Note that the reference contains an arithmetical error in the σ_b correction. CF_4 is given at 7.70 ppm downfield from liquid CF_3COOH (uncorrected).

[b]) Data for CF_4, SiF_4 and SF_6 at 35 °C relative to external liquid CF_3COOH; for C_6F_6 and p-$C_6H_4F_2$ measurements were made at 175 °C relative to internal CF_4 without correction for the temperature dependence of these shifts. Data from Raynes and Raza [103]. CF_4 is given as being 8.47 ppm downfield from liquid CF_3COOH (uncorrected).

[c]) At $t = 30\,°C$, internally referenced and extrapolated to zero pressure. CF_4 is 7.596 ppm downfield from liquid ethyltrifluoroacetate, uncorrected for the bulk susceptibility of the liquid. Data from Mohanty and Bernstein [52b]. Note that the reference contains an arithmetical error in the σ_b correction.

[d]) Data measured at 40 °C relative to external liquid $CFCl_2CFCl_2$; converted to gaseous CF_4 on the basis that gaseous CF_4 is 164.0 Hz (2.905 ppm) upfield from liquid $CFCl_2CFCl_2$

(uncorrected). Pressure was either 1–2 atm, or the vapour pressure at 40 °C for the less volatile compounds. No correction for the σ_b or σ_w shift of the vapour. Data from Abraham, Wileman and Bedford [98].

[e]) At 34 °C, internally referenced and extrapolated to zero pressure. Original data were relative to SiF_4; a correction of −104.18 ppm (see c) has been used to convert to the CF_4 base. Data of Hinderman and Cornwell [61].

A few concluding remarks should be made at this point. The σ_w effect for the ^{19}F is obviously large and rather variable (2–9 ppm). In comparison to protons, the ^{19}F σ_w effect and its range is about 20x larger. This makes it all the more important to know these effects, so that they properly be taken into account in discussions of intramolecular ^{19}F shifts. It is particularly important in ^{19}F NMR to recognise that internal referencing with variety of reference materials will give rather widely different shift data. The necessary corrections for virtually all the reference compounds ever used can be extracted from the work of Abraham *et al.* [98, 99]. In this regard it is interesting to observe that the ASTM recommendation [107] calls for an *external* reference, *viz* C_6F_6, as primary reference. Other reference materials (including internal ones) are permissible, but their relation to external C_6F_6 should be stated. This, of course, is only half the story; to interpret the ^{19}F chemical shifts of solutes in an intramolecular sense, suitable solvent effect corrections should be made. Unfortunately, at present, there is apparently no systematic method to estimate these, in spite of the availability of a rather large number of experimental σ_w data.

^{19}F has one other advantage relative to ^{1}H solvent effects. The neighbour anisotropy effect σ_a is independent of the nucleus studied; whereas in proton work σ_a is of the same order of magnitude as σ_w, in ^{19}F work the σ_a contribution is relatively minor and even the crudest sort of model for σ_a would suffice to make a suitable correction.

A study of ^{19}F Van der Waals effects may also be useful for the understanding of similar effects in ^{13}C, ^{29}Si, ^{31}P etc., whereas the same cannot be said of proton solvent effects; the nucleus ^{19}F is more typical because of the large contribution of the paramagnetic term.

Chapter 14. σ_w of Nuclei other than ^{1}H and ^{19}F

14.1. σ_w of ^{13}C and ^{29}Si

We discuss these two nuclei together because the only reports on pure Van der Waals solvent effects of these nuclei are from one source (*i.e.* Maciel and co-workers [100, 108, 109]), where these two resonances were studied simultaneously. We also will deviate from our general rule of discussing only gas shifts or gas-to-liquid shifts, because of the paucity of such data. The first report on σ_w of ^{13}C and ^{29}Si is due to Bacon, Maciel, Musker and Scholl [108], who for TMS in 20 vol % TMS, 80 vol % C_6H_{12} reported susceptibility corrected values, relative to pure TMS, of $\sigma_w(^{13}C) = +0.12$ ppm and $\sigma_w(^{29}Si) = -0.05$ ppm. In subsequent paper [109] several more of such data were given, but a full discussion of the ^{13}C, ^{29}Si and ^{1}H solvent effects of TMS and cyclohexane by Bacon and Maciel [*100*] includes all previous data. In Table 41 all the relevant data (*i.e.* pertaining to pure σ_w effects only) is given.

Table 41. $\Delta\sigma_w$ effects in ppm for ^{13}C, ^{29}Si and ^{1}H of TMS and C_6H_{12} according to Bacon and Maciel [*100*]. Data are corrected for susceptibility differences. Temperature was 40 °C. External reference was pure TMS for data on TMS, and 2% C_6H_{12} in TMS for the C_6H_{12} data. Solvent includes 20 vol % TMS both for TMS and C_6H_{12} (2% in 20% vol TMS, 78 vol % "solvent"). Data are re-referenced to $\sigma_w = 0$ for TMS or C_6H_{12} (^{1}H only) as solvent

solute	solvent	C_6H_{12}	TMS	CCl_4
TMS	^{1}H	−0.036	0.000	−0.16
	^{13}C	+0.12	0.000	−0.98
	^{29}Si	−0.05	0.000	−
C_6H_{12}	^{1}H	0.000	−	−0.14
	^{13}C	−0.02	−	−0.45

Not much can be derived from Table 41, except perhaps that the range for ^{13}C σ_w effects is much larger than for ^{1}H.

The study of Bacon and Maciel includes a great number of other solvents, such as aromatics, halo-cyclohexanes and halo-methanes. The authors have attempted a factor analysis on a group of data, comprising 38 solvents and the 5 resonances of TMS and C_6H_{12} as solutes. They claimed that two eigenvalues, i. e. two factors, were sufficient to cover the effects within experimental error for a limited set of 15 solvents, but that for the full set at least 3 vectors were necessary. They found that solvent polarity is probably *not* the third factor; C_6F_6 and CS_2 appeared to be the major carriers of the third solvent factor, which, *inter alia*, is interesting in view of our discussion of factor analysis (section 12.4). The authors did find a rotation for which the solute factors are nearly equal for the three nuclei of TMS. They identified the corresponding effect as a σ_a effect and then found it to correspond to about +0.35 ppm for benzene and +0.30 ppm for halo-benzenes. Rather then taking the accompanying vector as the σ_w vector, they carried out an independent search (i. e. rotation) for σ_w. The criteria they used are faulty, however. They state, for example, that the Van der Waals solvent factor of C_6H_6 as solvent should have a larger absolute magnitude than that of C_6H_{12}, and that similarly that of C_6H_5I (or other halo-benzenes) should be larger than that of C_6H_6. Neither of the models discussed in Chapters 2 and 3 allows such a simple statement. The authors, based on the above criteria, then find a "best" rotation, which turns out to be very different from the earlier rotations, found on the basis of "nearly-solute-independent-σ_a". The authors conclude with the truly astounding statement that "this result suggests a need for a re-examining of the premises on which existing pictures of solvent effect contributions, particularly dispersion shifts, are based". Not only would it be more logical to doubt the premises accepted by the authors, but more fundamentally, such a conclusion cannot be arrived at at all; Mathematics can never make a statement regarding Physics!

14.2. σ_w of ^{31}P

^{31}P NMR has been increasingly popular if only because of the tremendous potential
in biochemical applications. Surprisingly enough there is very little information on
^{31}P solvent effects. Yet, from the information there is it seems clear that in ^{31}P NMR
the solvent effects are very large, about 100 times larger than with protons and about
5 times larger than with ^{19}F. It would seem thus doubly necessary to know these
effects.

Heckmann and Fluck [110] have measured elemental phosphorus P_4 in the gas
phase. The vapour was studied while in equilibrium with liquid (white) phosphorus.
By varying the temperature the density of the vapour could be changed. The liquid
was also measured at the same time, all referenced to external 85% aqueous H_3PO_4
solution; by a separate experiment it had been established that the temperature
dependence of the resonance frequency of the phosphoric acid solution was negligible.
In Table 42 their results for liquid and gaseous phosphorus are given. The data for
the liquid have been corrected for the bulk susceptibility of the H_3PO_4 solution
(1.58 ppm) and of the liquid phosphorus (3.20 ppm). Changes of this contribution
with temperature, as well as the σ_b correction for the vapour have been neglected.

Table 42. Relative chemical shifts, corrected for bulk susceptibility, upfield from 85% aqueous H_3PO_4 as external standard, of liquid and gaseous elemental phosphorus according
to Heckmann and Fluck [110]

temp (°C)	vapour pressure	$\delta_m - \delta_b$ (ppm) liquid phosphorus	δ (ppm) P_4 vapour
253	440	476.0	549.9
265	560	476.8	549.7
280	760	477.9	548.7
295	990	479.0	548.5

The vapour data must be considered as due to a composite density and temperature
effect. An approximate separation may be made as follows; the vapour shifts of
Table 42 extrapolated to zero pressure give 551.5 ppm, which value we shall also
use at $t = 30\,°C$. This extrapolation is not quite correct, but the error will be small.
From the data for liquid phosphorus at lower temperature given also by the same
authors [111] we deduce a corrected shift $\delta_w = \delta - \delta_b$ of 461.7 ppm at $30\,°C$. There
is thus a Van der Waals shift of $\sigma_w = -89.8$ ppm (this appears in variance with the
value of -93.0 ppm quoted by the authors [110] but the latter datum was apparently
not corrected for σ_b). With a density of $\rho = 1.82$ and a molecular weight of 124 (for
P_4) one finds

$$\sigma_1\ (t = 30\,°C) = \Delta\sigma_w/\Delta(1/V_M) = -5955 \text{ ppm cm}^3 \text{ mole}^{-1}$$

If next this σ_1 value is also used for the vapour data, ignoring the fact that at these
higher temperatures σ_1 will be somewhat different, one can estimate $(d\sigma/dt)_V$ from

$$\Delta\sigma = \left(\frac{d\sigma}{d\,1/V_M}\right)_t \Delta(1/V_M) + \left(\frac{d\sigma}{dt}\right)_V \Delta t \tag{156}$$

Using the data from Table 42 one then finds that the density term is almost negligible compared to $\Delta\sigma = -1.4$ ppm. One finds $(d\sigma/dt)_V = -3.1$ ppm / 100 °C versus -3.3 ppm / 100 °C for the total effect. In the liquid a positive temperature dependence on σ is found, no doubt because here the density effect outweighs the intra- and intermolecular temperature effects. The authors give the temperature effect on liquid phosphorus for 0 to 295 °C [110, 111]. There are apparently two linear regions; from 60 to 140 °C one has $\Delta\sigma/\Delta t = +4.3$ ppm / 100 °C while from 190 to 300 °C one has $\Delta\sigma/\Delta t = +7.1$ ppm / 100 °C. Below 60 °C the temperature dependency shows another irregularity, particularly around the melting point. In Table 43 the available data on P_4 in non-polar solvents are given. The gas-to-liquid σ_w data have been calculated on the basis of 551.5 ppm of P_4 vapour quoted above. A correction of 0.5 ppm has been applied on the gas-to-benzene shift to account for the anisotropy effect σ_a.

Table 43. Chemical shifts (relative to 85% aqueous H_3PO_4) and Van der Waals shifts of P_4 in non-polar solvents, according to Heckmann and Fluck [110, 111, 112]. Temperature 30 °C

solvent	δ (ppm) uncorrected	δ (ppm) corrected	$-\sigma_w$ (ppm)	ref
$n-C_5H_{12}$	530.9	530.4	21.1	112
C_6H_{12}	528.3	528.0	23.5	112
C_6H_6	522.3	522.0	30.0	111
CCl_4	521	521	30.5	111
CS_2	511.3	511.1	40.4	111
phosphor (white)	460.1	461.7	89.8	111

The authors have also tried to give a physical picture of these solvent shifts, but they use for this purpose only the solvent dependence according to McRae's formula [see Eq. (36)] and the even more primitive model of Glick, Kates and Ehrenson [6] (see Chapter 1). Rather than concluding that these models are inadequate, they continue by assuming that these models give the correct values for the diamagnetic dispersive contribution to the solvent shift, and by attributing the deviations to the diamagnetic part of the repulsive contribution to these same shifts. The claims some people manage to get past unsuspecting referees...!

14.3. σ_w of ^{129}Xe

Nearly all the available information on σ_w of ^{129}Xe was already discussed in foregoing pages, notably in sections 7.3, 9.2, 9.6, chapter 10 and section 11.1. The experimental $(\sigma_1)_w$ data on ^{129}Xe are given below in Table 44.

Table 44. Experimental $(\sigma_1)_w$ data for ^{129}Xe, in various solvent gases. Data were obtained at 25 °C and were corrected for bulk susceptibility

solvent	$\dfrac{-(\sigma_1)_w}{\text{ppm cm}^3 \text{ mole}^{-1}}$	remarks
Ar	3,080	[66]
CO_2	3,779	[66]
CF_4	4,262	[66]
CH_4	6,165	[66]
Kr	6,009	[66]
Xe	12,188	[66] also $(\sigma_2)_w$, $(\sigma_3)_w$
	9,600	[78]
	9,360	[79] gas *and* liquid Xe
	10,280	[113]
	11,800	[114] solid Xe
	13,000	[80] also $(\sigma_2)_w$ given
O_2	28,015	[115]
NO	20,736	[115]

The last two entries of Table 44 due to Jameson and Jameson [*115*] are very surprising; for perturbers like O_2 and NO one would expect a $(\sigma_1)_w$ of about -3000 ppm cm^3 mole $^{-1}$, even after the large upfield correction for $(\sigma_1)_b$ of these paramagnetic perturbers. Buckingham and Kollman [*116*] have given the explanation for this phenomenon. They show that it cannot be due to a "pseudo-contact" contribution to the shielding. In that case the effect should be very much larger for NO than for O_2 while furthermore some simple calculations show that even for NO the "pseudo-contract" contribution is too small by two orders of magnitude. Buckingham and Kollman then show rather convincingly that the effect is due to a direct contact mechanism; *i.e.* a non-zero unpaired electron density at the nucleus of the perturbed atom, coupled with the preferred β-orientation in the magnetic field. The authors could not quite quantitatively calculate and reproduce the experimental data, mainly due to the incomplete knowledge of the Xenon wave functions. The effect is also very sensitive to the intermolecular potential, so that in principle this contact solvent effect should be a sensitive tool for the study of intermolecular potentials.

Chapter 15. Alternate Referencing Systems

15.1. Internally Referenced σ_w Data

With very few exceptions we have concerned ourselves in the foregoing with externally referenced measurements, either in the gas phase or in the liquid phase (solution phase). The reasons for our evident disdain for internally referenced studies should be obvious; the results so obtained are always the *difference* between two σ_w effects; that of the solute and that of the reference compound. In general, solute and reference will exhibit σ_w effects in various solvents, which are not related in a simple way. This is even true if the "reference" is a chosen signal of the solute molecule itself; as we have seen (Chapter 4 section 6.1), *intra*molecular differences in σ_w can be just as substantial as *inter*molecular differences.

Nevertheless there are substantial advantages to internal referencing. For example, the experimental technique is much easier and faster. Also, the disadvantages may be minimised; if a number of solutes are investigated in the same solvent and with the same internal reference, all the solvent effects so obtained are in "error" by the same constant amount. For purposes of finding or utilising certain trends this is no great disadvantage, but one should realise that always one basic bit of information is lost (i.e. the *total* medium shift of solute X), while another bit of non-interesting information (solvent effect of the reference) is added. Figure 9 indicates how inter-

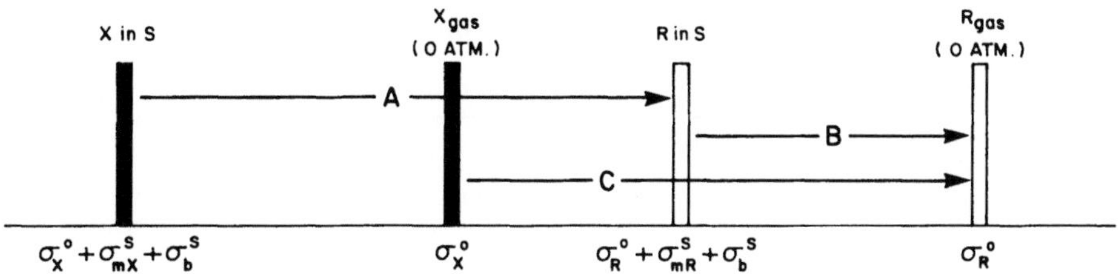

Fig. 9. Schematic conversion of internally referenced shift data of solute X in solvent S with reference compound R into gas-to-liquid data: $A + B - C = -(\sigma^S_{mX} + \sigma^S_b)$

nally referenced data can be converted to true gas-to-liquid shifts. Apart from the internal shift (A) two extra data are required. Firstly the uncorrected gas-to-solution shift (B) of the reference compound. Many of these data are available such as from the Tables in this review, supplemented with calculated σ_b contributions. This measurement is usually obtained in two separate steps where R (gas) and R (in solvent S) are measured separately relative to an intermediate external standard (such as capillary with acetone). Secondly, the gas phase difference (C) is required. This can be done by preparing a mixture of the two compounds so that each will have about 1 atm of partial pressure. This is then actually an internally referenced shift, but at 1 atm pressure the difference in medium effects is usually less than 0.001 ppm. The procedure usually requires heating of the sample to produce sufficient vapour pressure. In fact, the measurements should be done at several temperatures to "eliminate" intramolecular temperature effects (see section 11.1). Fourier Transform Techniques may prove of substantial use here in enabling the measurement of gas phase spectra of compounds of low volatility.

From the three pieces of information, the gas-to-liquid shift $A + B - C = -(\sigma^S_{mX} + \sigma^S_b)$ can now easily be found. A variation of the above has been given by Becconsall *et al.* [117]. It is again a difference method, but rather than giving the difference between the two solutes in one solvent, it produces differences between the σ_w's (or in general the σ_m values) of one solute in two solvents. Although this kind of information is of importance particularly if the two solvents are very different in nature (for example one being non-polar isotropic such as CCl_4 and the other being anisotropic such as C_6H_6 or pyridine) the method is also of potential value for studying pure Van der Waals solvents. Consider the difference $\Delta^{x,y}$ between the *internally* referenced chemical shifts of solute S in two solvents x and y. One has then, if R is the reference compound,

$$\Delta^{x,y} = \delta(y) - \delta(x) = \sigma_{mS}(y) - \sigma_{mS}(x) - \sigma_{mR}(y) + \sigma_{mR}(x) \tag{157}$$

As foreseen, the above $\Delta^{x,y}$ still contains the medium effects of the reference. Becconsall *et al.* proposed to eliminate the latter by measuring the reference compound in the two solvents, once in a conventional magnet system (where the sample is perpendicular to the main field H_0) followed by a measurement of the same sample but now with a superconducting magnet (where the sample is parallel to the main field H_0). Note that these measurements have to be done with an external reference, but upon taking the difference of R in two solvents, this intermediate external reference is again eliminated. Since the bulk susceptibility depends on geometry

$$\sigma_{b\perp} = \frac{2\pi}{3}\chi_v \text{ and } \sigma_{b\parallel} = -\frac{4\pi}{3}\chi_v \tag{158}$$

one has

$$\sigma_{R\perp}(y) - \sigma_{R\perp}(x) = \frac{2\pi}{3}\{\chi_v(y) - \chi_v(x)\} + \sigma_{mR}(y) - \sigma_{mR}(x) \tag{159}$$

$$\sigma_{R\parallel}(y) - \sigma_{R\parallel}(x) = \frac{-4\pi}{3}\{\chi_v(y) - \chi_v(x)\} + \sigma_{mR}(y) - \sigma_{mR}(x) \tag{160}$$

Upon combining the above one finds

$$\sigma_{mR}(y) - \sigma_{mR}(x) = \frac{1}{3}[\{\sigma_{R\parallel}(y) - \sigma_{R\parallel}(x)\} + 2\{\sigma_{R\perp}(y) - \sigma_{R\perp}(x)\}] \tag{161}$$

For any chosen reference, such as TMS, this correction need therefore be determined only once for any solvent pair. after which a reference-free medium shift difference $\overline{\Delta}^{x,y}$ can be obtained from

$$\overline{\Delta}^{x,y} = \Delta^{x,y} + \sigma_{mR}(y) - \sigma_{mR}(x) \tag{162}$$

The above defined $\overline{\Delta}^{x,y}$ has also been used by Krystynak and Rummens [*118*]; however, these authors determined the correction $\sigma_{mR}(y) - \sigma_{mR}(x)$ experimentally, including a σ_b correction based on literature χ_M data. The two methods are in principle equivalent, but the technique of Becconsall *et al.* is more accurate, because it avoids possible systematic errors due to uncertainty in χ_M data. It should be stressed, however, that with the Becconsall method one obtains only *differences*; to obtain *total* medium shifts it is still necessary that for each solute the gas-to-liquid shift be measured in at least *one* solvent. Such a measurement would again involve a literature χ_M datum, the uncertainty of which would be reflected in all the gas-to-liquid data based upon this technique resulting in a constant systematic error. A disadvantage of the technique of Becconsall *et al.* is that it requires the probes of the two magnets to be at exactly the same temperature.

100

15.2. Externally Referenced Measurements

We are referring here to solvent effect data referenced to an external reference, either with or without correction for the bulk susceptibilities of one or both the liquids, but without measurement of the solute X in the gas phase. Such external referencing is recommended when the solvent study in question deals with more than one solvent. In such case internal referencing is of very limited use and may even lead to grossly erroneous conclusions [118]. The advantage of external referencing is, of course, that the influence of the reference is limited to a constant term for all the solute-solvent systems studied. If more than one solute is studied, much of this advantage is dissipated, because now the inter-solute comparisons include the solute gas phase shift differences as extra terms. The conversion to gas-to-liquid shifts is easier than with internal referencing. In addition to gas phase measurements for all solutes relative to the reference compound in the gas phase, only the one-time measurement gas-to-liquid of the reference compound is required. Figure 10 shows schematically how such a conversion may be achieved. It may easily be verified that the three experi-

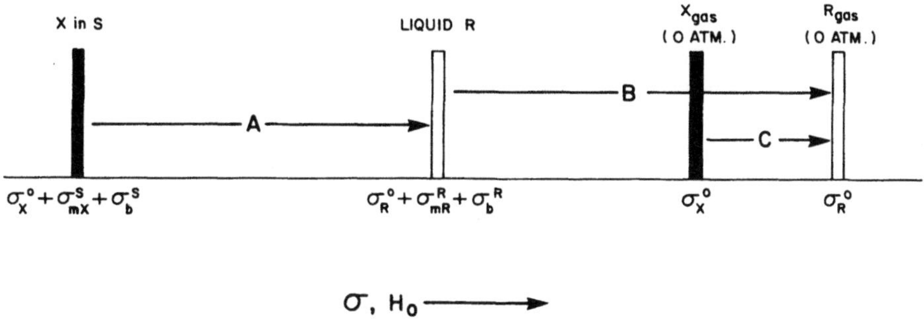

Fig. 10. Schematic conversion of externally referenced shift data of solute X in solvent S relative to liquid reference R into gas-to-liquid data; $A + B - C = -(\sigma_{mX}^S + \sigma_b^S)$

mental data A, B and C may be combined to give the downfield gas-to-liquid shift of the solute X; $A + B - C = -(\sigma_{mX}^S + \sigma_b^S)$. It is interesting to notice that while the initial externally referenced shift (A) contains the difference of two susceptibility terms, the susceptibility of the reference liquid need not be known to find the gas-to-liquid shift of X.

External referencing is not an easy technique. Apart from the uncertainties in χ_M, the σ_b correction harbours a potential for many systematic errors. So that one may use literature values of χ_M, the solutions should be rigorously free of paramagnetic and ferromagnetic impurities. This refers not only to dissolved oxygen (whose contribution is rarely more than 0.01 to 0.02 ppm), but experience has taught us that virtually all laboratory equipment and every chemical is contaminated with measurable amounts of paramagnetics. The reference to ferromagnetics is no joke either; Broersma already reported [20] that he had to devise an electromagnetic filter as a final recourse to obtain reproducible susceptibility data. The liquids have also to be chemically pure. Apart from non-intentioned impurities, finite concentrations of the solute (and of the reference if the initial measurement was internally referenced) may substantially change the volume susceptibility. Even at 1% solute concentration the error may amount to 0.02 ppm. One may make a correction for this by assuming that susceptibilities are additive on a volume % basis, but in principle *and* in practice it is better to measure the shifts at a number of concentrations and extrapolate the results to zero solute concentration. These plots are doubly useful in that they instill the proper attitude of modesty into the novice experimenter (the plots will look hopeless the first few weeks of trying) as well as a feeling of reassurance and confidence once they look nice without more than ±0.002 ppm scatter, which is achievable after the required magic touch, the "Fingerspitzengefühl", has been developed. The geometry of the sample is equally important; it is absolutely mandatory that high precision sample tubes be used with a minimal degree of conical or ellipsal deformations. Reference capillaries must be carefully centered. For capillaries in particular it is difficult to fulfil the cylindricity requirements, unless one buys ultra-high precision tubing. In one illuminating experiment the author once prepared a dozen acetone-filled capillaries, made from melting point capillaries; the observed range of resonances was 0.08 ppm! The author presently prefers the coaxial cells manufactured by Wilmad [119]; the annular space, to be filled with reference liquid, is so thin that the liquid capillary forces act as a self-centering mechanism; nevertheless repositioning of the two tubes may lead to observed shift changes of ±0.002 ppm. An excellent discussion of the above matters has been given by Zimmerman and Foster [120], who also consider the effects of glass wall roughness, of rotation and of non-perpendicularity with the magnetic field.

Most of the above indicated problems are compounded if the compounds to be used are volatile or sensitive to O_2 or H_2O.

15.3. χ_M and the Bulk Susceptibility Correction σ_b

The topic of this section would be quite worthy of a review the size of the present one. Without even attempting to be comprehensive and complete we think it nevertheless useful to make a number of comments.

In many of the existing NMR solvent studies, use has been made of χ_M data compiled by Smith [121]. This compilation is incomplete, uncritical and contains many poor data. The very existence of this compilation leads one to wonder, because at that time there existed already (since 1957) a very complete and critically reviewed compilation [21]. The French title of this work may have frightened some, but it is remarkable how even a IUPAC sponsorhip and a multi-language introduction could not prevent it from being ignored. A more recent compilation (1967) under the Landolt-Börnstein aegis [22] is now available, but even this has not moved the Chemical Rubber Company and neither any scientists, from still today only quoting the Smith compilation as given in the much used "Handbook of Chemistry and Physics". May we gently plead that American Data are not always necessarily Superior?

Our second remark is deemed to be considered trivial. The σ_b correction is proportional to χ_v, not just to χ_M, so that it is a function of density ρ, and therefore of temperature. Most published density data refer to $t = 20$ °C, but most NMR spectrometers operate at 25–40 °C. The uncritical use of ρ^{20} data for the σ_b correction may thus lead to gross errors, of up to 0.1 ppm. In spite of this rather obvious fact, many of the studies quoted in this review suffer from this deficiency; where possible we have made the necessary correction, but often even the probe temperature was not stated.

The third remark is aimed at emphasising that the factor $2\pi/3$ in the expression $\sigma_b = 2\pi\chi_v/3$ is for an *infinitely* long cylinder. Dickinson [1], using a 10 mm insert, with 8 mm OD, 5.4 mm ID sample tubes, measured the susceptibility factor, using total sample lengths of 5, 10, 20 and 100 mm. At 20 mm (i.e. a ratio 2 : 1 of sample length to insert coil diameter) the deviation from the limiting value was about 12%. He considered the limiting value being obtained at 100 mm (i.e. a length-to-coil diameter ratio of 10 : 1), but data at intermediate lengths are lacking. Lussan [122], using a standard 5 mm probe with a coaxial sample cell and measuring ethanol shifts as function of cobaltion concentration, found that the factor $2\pi/3 = 2.094$ was correct within ±0.02. The present author, using also a 5 mm probe with a total liquid sample length of 50 mm has found that shifts, either relative to a standard in a centred narrow capillary or relative to the same standard in the annular space of a Wilmad coaxial cell, were not systematically different. It appears therefore that in a 5 mm probe a sample length of 50 mm is quite adequate; whether this length is also sufficient in 10 or 15 mm probes remains an open question.

Susceptibilities are preferably measured by classical balance methods of Pascal and of Gouy. But given a number of absolute and accurate χ_M data, NMR methods can be used in a variety of ways. Frei and Bernstein [123], for example, inserted both a cylindrical and spherical reference cell inside a standard 5 mm sample tube. The frequencies of the reference compound in the two cells will be different by an amount

$$\delta_{cyl} - \delta_{sphere} = (\alpha_{cyl} - \alpha_{sphere})\,(\chi_v(\text{ref}) - \chi_v(\text{sample})) \tag{163}$$

Ideally the difference between the two shape factors α should be $2\pi/3 = 2.094$, but calibration with 15 samples of accurately known χ_v produced a slope of 2.05. This deviation is probably due to the non-ideal geometry (the sphere required a supporting stem). Nevertheless, once given the calibration constant, relative χ_v data can be obtained with a precision of about ±0.2%. Vincent *et al.* [124] made a virtue out of the necessity of supporting the sphere, by filling the stem also with liquid and by moving first the sphere, then the stem into the r.f. coil region or even observing the two signals simultaneously. They claim a precision of ±0.3% for their relative measurements of χ_v.

Zimmerman and Foster [120] have described the field conditions in a non-spinning coaxial cell. Numbering the five regions 1 to 5 from the inside and describing the position of a point anywhere in the cell by polar coordinates r and θ (relative to the center of the cell and to the direction of H_0) they find for the field at a point in the third, annular, region

$$\frac{H_3 - \langle H_3\rangle}{H_0} = -\frac{2\pi a_1^2(\chi_2 - \chi_1) + a_2^2(\chi_3 - \chi_2)\cos 2\theta}{3\,r^3} \tag{164}$$

where $\langle H_3 \rangle$ is the average field in the annular space (observed when spinning) while a_1 and a_2 are the radii of the outer boundaries for the regions 1 and 2. Note; this formula was also derived by Morin *et al.* [*125*]. The result of the field distribution as described by Eq. (164) is a signal shape which is broad, but with two sharp maxima, symmetrically displaced around the centre. Note that the inner tube might also be a solid rod of glass so that the term $(\chi_2 - \chi_1)$ then vanishes. Also for a sufficiently thin annular space r is virtually constant and equal to a_2. Under these conditions the signal width, and also the distance between the two maxima of the signal becomes proportional to $(\chi_3 - \chi_2)$. Douglas and Fratiello [*126*] have used this technique to measure χ_v, after suitable calibration. The method is fast, but not quite as precise as other methods.

When a coaxial cell is spun, the average field in the region i is given [*120*] by

$$\langle H_i \rangle / H_0 = 1 - \frac{2\pi}{3} (\chi_i - \chi_5) \tag{165}$$

where χ_5 is normally the susceptibility of air. For the difference between the central and the annular region one therefore has

$$\frac{\langle H_1 \rangle - \langle H_3 \rangle}{H_0} = \frac{2\pi}{3} (\chi_3 - \chi_1) \tag{166}$$

By keeping a reference compound in one region one can follow the changes in susceptibility in the other region, provided of course that whatever made the susceptibility change, does not at the same time change other contributions to the shielding. This can be achieved with paramagnetic ions at low concentration; Eq. (166) forms the basis of the above quoted work of Dickinson [*1*] and of Lussan [*122*].

A spinning coaxial system can also more directly be used to measure susceptibilities. It is well known by all who have ever used coaxial cells that the spinning side bands of the signals from the annular region are unusually strong. It is to the credit of Malinowski and co-workers [*127*] for having recognised that the high intensity of these side bands is not due to imperfections of the cell but rather due to a modulation effect. As a volume element in the annular region is rotated it goes periodically through a range of fields as predicted by Eq. (164). But since H_3 is periodic with π [because of the cos 2θ in Eq. (164)], the separation between two side bands ν_m should be twice the rotating frequency ν_s, as was indeed verified [*127*]. The authors then used the NMR modulation theory developed by Williams and Gutowsky [*128*] which predicts that the intensity of the n-th band is proportional to $J_n^2(k)$, where $J_n(k)$ is the n-th order Bessel function of the first kind with argument $k = \gamma \Delta H / 2\pi\nu_m$. By adjusting the spinning speed so that the first side band is equal in intensity to the main band one has $J_0^2(k) = J_1^2(k)$ which is true for $k = 1.4347$. Thus ΔH (the right hand side of Eq. (164) devided by cos 2θ) can be found, whence relative susceptibilities. Using two liquid systems the authors quote -0.817 and $-0.818 \cdot 10^{-6}$ emu for the susceptibility of the inner glass tube. Apparently the method is quite precise. Malinowski and Pierpaoli also showed [*129*] that field gradients can cause strong side bands with spacings of either $+n\nu_s$ or $-n\nu_s$. They appear only at one side of the main band however. With coaxial cells there is a natural field gradient ΔH as described above. In a manner entirely analogous to that described above, these gradient side bands can be used to measure susceptibility differences. One interesting difference is, however, that the gradient method gives an absolute sign of ΔH (and hence of $\Delta\chi$) depending on whether the side bands appear to the left or to the right of the main band.

A quite different method is based on the ideas of Becconsall *et al.* [*117*] to utilise the different shape factors for a cylinder in a conventional and in a superconducting magnet (see section 15.1). Whereas Becconsall *et al.* used this to eliminate the effect of the internal reference on medium effect differences, it is obvious that the same idea can also directly be used to measure χ_v. Subtraction of Eqs. (159) and (160) gives

$$2\pi[\chi_v(y) - \chi_v(x)] = [\sigma_{R\perp}(y) - \sigma_{R\perp}(x)] - [\sigma_{R\parallel}(y) - \sigma_{R\parallel}(x)] \tag{167}$$

This method has been studied by Homer and Whitney [*130*], who found it to be as accurate as can be hoped ($\pm 0.2\%$). Note that no calibration is required if precision coaxial tubes are used. Interestingly, the authors found that the finite solute concentration (0.005 mole % cyclohexane) was of no consequence on the obtained results. They also found that a temperature difference of 14 °C between the two probes gave virtually the same results for the *specific* susceptibilities provided densities appropriate to the probe temperatures were used! In both cases there is apparently a good cancellation of small differences in medium effects.

Finally we want to mention the rather exhaustive study on the effect of cell imperfections on χ_v determinations made by Frost and Hall [*131*]. They describe mostly the method of external reference using a small stem-supported sphere, but their considerations are useful for other systems too. Their main conclusion is that it is not possible to use a single shape factor to describe a non-perfect cell; it depends on the type of experiment performed what shape factor will result. This is caused by the fact that in non-perfect systems the susceptibility of the glass enters into the picture. The authors find, however, that any calibration technique with known susceptibilities will still be linear and will therefore always give reasonable results. They indicate that to minimise the effect of imperfect cell geometry, the solution under study should be in the sphere, with the reference liquid in the outer region and chosen in such a manner that its susceptibility matches that of the glass used for the sphere. They claim that it should then be possible to make a spherical cell which under these circumstances will have a shape factor exactly equal to zero. The authors also make an interesting remark in warning that the insertion of a sample into the magnet gap actually changes the outside field H_0 in a manner dependent on the susceptibilities of the sample cell and its contents. This causes problems if one uses a substitution method for measuring shifts. Substitution methods may seem oldfashioned, but they have recently undergone a revival of sorts since the introduction of super-stable magnet fields!

Chapter 16. On the Required Molecular Parameters and Physical Constants

16.1. Bulk Properties

In most σ_w studies it is important to know exactly the temperature at which one has been working. This is so, not only because σ_w itself is temperature dependent, but certainly because the susceptibility correction σ_b is strongly dependent on temperature (see section 15.3). It is possible to measure sample temperature by dipping a thermocouple or thermistor into the sample; for solvent effect studies with external reference where freedom from oxygen and other paramagnetic impurities is important, this method is not recommended. For gases and for volatile liquids this method is even impossible, because it requires an open-ended sample tube. The above problems may be circumvented by substituting the sample of interest by a sample tube, containing a thermocouple immersed in a liquid, and by measuring the temperature before and after the NMR measurement of interest. Even this technique is not without systematic errors; one will find that the temperature measured tends to depend on the amount of liquid in the thermocouple tube and on whether this tube is spun or not. The likely causes of these effects are temperature gradients over the sample (if the dewared insert is not of a countercurrent design) or thermal leakages through the thermocouple itself. With regard to the latter it must be remembered that the heat transfer via the thermostatting gas is rather inefficient; if the sample is a gas this problem is even augmented and the highly conducting thermocouple will act as a heat sink or heat source. Perhaps the best recommendation is to use the substitution method with carefully degassed samples of methanol (for low temperatures) or glycol (for high temperatures). The chemical shift of the hydroxyl groups measured relative to the alkyl groups of these liquids is strongly dependent upon temperature. A most careful calibration of this chemical shift thermometer has been carried out by Van Geet [132]. For methanol over the range 175–330 °K he finds within ±0.8 °K:

$$T(^\circ K) = 403.0 - 0.491\,\Delta\nu - 66.2 \cdot 10^{-4}(\Delta\nu)^2 \tag{168}$$

and for glycol over the range 310–410 °K with a precision of ±0.3 °K

$$T(^\circ K) = 466.0 - 1.694\,\Delta\nu \tag{169}$$

where $\Delta\nu$ is expressed in Hz at 60.0 MHz.

We would also like to remark that in most probe and insert designs only the lower part of the sample tube is thermostatted, whereas the upper end usually has a temperature, close to the ambient probe temperature. One may deal with this problem by making an extra internal seal in the sample tube about 6 cm above the bottom as described by Rummens [24]. On the other hand, for the most precise work one may want to insist on a refillable single sample tube for the entire study where such a seal is not possible. Such a system has been described by Mohanty and Bernstein [52b] for their studies of ^{19}F σ_w in gases, but now the above noted problem of a large temperature gradient over the entire sample tube enters again. Would not the density of the gas in the thermostatted region near the r.f. coil be significantly different from the overall average density, the latter being the density used in the determination of σ_1?

With the above question we have entered already the discussion of the second important bulk property *viz* the density ρ. For liquids the required density *versus* temperature information may be found in a variety of standard references [*133*], but for gases the density has to be determined for each experiment. A most versatile system for condensable gases has been described by Mohanty and Bernstein [*52b*]. The vapours are condensed, either totally or partially, from an attached reservoir, or rather a set of reservoirs, whose volumes are exactly known. The amount of gas condensed can be determined from the reservoir volume and the drop of pressure in the reservoir, the latter being measured usually by a mercury manometer. By choosing an appropriate combination of reservoir volume and pressure drop, an accurate determination of the amount condensed (±0.2%) can be obtained. Since the gas pressures measured are low (sub-atmospheric), it is usually permissible to assume ideal gas behaviour; for certain gases a fugacity correction may be needed, however. To calculate the average gas density in the sample tube it is additionally necessary to know the sample tube volume; with accurate calibration, particularly with a valve-operated sample tube [*52b*], the tube volume may be determined with an accuracy of ±0.1% in all but the smallest or narrowest tubes. For non-condensable gases, a high pressure sampling device is required; the use of a valve-operated, de-connectable sample tube is then almost mandatory; the density must then be determined from measured pressures and known p, V, T data of the gas.

For a discussion of a third important bulk property, *viz* the refractive index, see the following section.

16.2. Molecular Properties

In the calculation of σ_w with whatever model, one requires a number of molecular parameters. It is of some importance to agree on certain conventions here, so that calculated σ_w data be intercomparable.

The molar susceptibility χ_M was discussed already in section 15.3, whereas the molar volume V_M needs no further discussion after the comments made on density in the previous section.

As far as the molar polarisability α is concerned, we would like to propose that this quantity be defined by the Lorentz-Lorenz equation [Eq. (27)]. This is not quite correct, since the Lorentz internal field is not correct, certainly not for liquids. The Onsager field is much better, but one has then the difficulty of having to determine simultaneously the cavity radius a. The Onsager relation $4\pi a^3 N = 3 V_M$ [Eq. (23)] is of no use, since this assumption will reduce the Onsager theory for non-polar compounds to the Lorentz-Lorenz equation once more. The latter is easy to use, however, since refractive indices of many compounds are available [*133*]. The only precaution needed here is to use density data referring to the same temperature as for which the refractive indices are given (which need not be the temperature of the NMR experiment). The α's so determined ought to be multiplied by $(1 - g\alpha)$ to obtain the presumably better Onsager polarisabilities. For non-polar compounds this factor $(1 - g\alpha)$ varies between 0.8 and 0.9; omission results therefore in α's which are about 15% too large. However, such an almost constant factor is easily taken up by the calibration procedure for σ_w; in the binary gas model for example this is incorporated into the systematic uncertainty of the B parameter. Refractive indices are usually given at the sodium D-line wavelength. To be exact one ought to extrapolate to infinite wavelength using, for example, the Drude dispersion theory. Again, for most compounds this correction is close to constant, the D-line based polarisabilities being a few % too large, which once more is incorporated in the calibration procedure. For coloured compounds, or more precisely those whose high refractive index at the sodium D-line wavelength is not due to the presumed resonance transition (but for example to some partially forbidden low energy transition), an extrapolation to infinite wavelength may be necessary.

Most theories and models for σ_w end up by applying the London approximation $h\nu \approx I$, the error induced being considered as not too large and as being systematic and nearly constant. Ionisation potentials have been compiled by Kiser [134]. This list, while being relatively short, is presumably good enough. Ionisation potentials are virtually constant over any class of chemical compounds, the entire range is relatively small (about a factor of two) and unknown ionisation potentials can usually be estimated with good precision ($\pm 5\%$) from those of related compounds. Resonant frequencies are medium dependent; whereas in section 2.3 we have criticised the use of the McRae-Bayliss theory on NMR transitions, that theory is fully applicable to the "resonant frequencies" which occur in all σ_w models. Therefore these frequencies, or even the ionisation potentials, might be corrected for this medium effect, but such a proposal we have not yet seen studied.

In the calculation of site factors (section 6.1) the parameter d is required, which is defined as the distance from nuclear site of interest to the centre of mass of the solute molecule. Since this parameter is exclusively dependent upon the structure of the solute molecule, its determination usually poses no special problems. However, with molecules possessing internal motions some averaging has to be carried out. For example in cyclohexane the d parameters for axial and equatorial protons are different; a simple average value for d will suffice here. In tetraethylsilane the methyl rotation causes $d\,(CH_3)$ to be conformation dependent. The author in an earlier paper [17] took the simple average of the maximum and the minimum value for d, but this is not quite correct. It is better to assume a preferred conformation, in this case with the methyl group staggered relative to the CH_2 group and to then calculate the average d for the *three* protons of the methyl group [57, 58]. A similar procedure could be used for hexyne-3. In both examples, however, there is a further complicating factor in that the ethyl groups are only equivalent on a long time scale; at each individual moment they can be non-equivalent (in tetraethylsilane at most three ethyl groups can be equivalent at a given time). This instantaneous non-equivalence causes the centre of mass to deplace itself; the d parameter of any group thus becomes a function of the conformations of the other groups. The effect is usually small and little accuracy is lost by ignoring it. With other molecules it is necessary to know the exact conformational preference. In *trans*-butene-2 for example each of the methyl groups has one C H bond which is in-plane with the double bond and pointing to this double bond. If another conformation were assumed, say by rotation of 180° relative to the one discussed above, a very different d parameter would result.

16.3. The Lennard-Jones (6–12) Force Constants ϵ/k and r_0 and the Functions $\mathscr{H}_6(y)$ and S_6^g

Throughout this review the author has maintained a barely disguised bias towards the binary collision statistical model for the calculation of σ_w. Indeed, for all its shortcomings, no other model can approach the universality, the flexibility and the kind of quantitative agreement with experiment that this model affords. What then is the reason that so few others have used this model and have instead preferred to use other, less accurate models? The answer is probably that the statistical model requires at least two extra parameters, *viz* those to describe the potential. These parameters can be obtained from measurements of diffusion coefficients, second virial coefficients and viscosities of gases, but the number of known parameters is limited, while those that are known show sometimes a disquieting divergence. The most complete compilation has been given by Hirschfelder, Curtis and Bird [135]. We will cite here only one example of the kind of differences one may encounter, *viz* that of C_2H_6 which case was already partly indicated in the fourth entry of Table 25 (section 9.6). The viscosity-based parameters for C_2H_6 are $\epsilon/k = 230$ and $r_0 = 4.418$, while those based on the second virial coefficient B(T) of the equation of state are $\epsilon/k = 243$ and $r_0 = 3.954$ [135]. Looking at the impact these differences have on the calculated σ_w (Eq. 76), we calculate at $T = 300\ °K$ $\mathscr{H}_6(y)/y^4 r_0^3 = 0.0772$ and $S_6^g = 1.545$ for the first set and $\mathscr{H}_6(y)/y^4 r_0^3 = 0.1103$ and $S_6^g = 1.742$ for the second set. We see thus that the two σ_w calculations will differ by a factor of 1.61. One also notices that uncertainties in r_0^{-3} are amplified by the site factor. In fact, not only in this particular case, but almost universally so will the differences in ϵ/k and in r_0 be such that they amplify each other in the $\mathscr{H}_6(y)/y^4 r_0^3$ term. Hirschfelder *et al.* state [135] that for transport properties one should use η-based parameters, while for steady-state properties (such as σ_w!) the $B(T)$ parameters ought to be better. We are in agreement with this principle, but believe that in practice this is only correct for small molecules. For larger molecules ϵ/k is also larger so that in general the reduced temperature $T^* = kT/\epsilon$ is smaller. Most of the $B(T)$ measurements on larger molecules are therefore done in a region where the reduced second virial coefficient $B^*(T^*)$ is almost linear in T^*. Not only is now the extraction of ϵ/k and r_0 rather insensitive, but the derived temperature dependence of $B(T)$ becomes highly susceptible to whether or not higher virial coefficients were included in the treatment. This is even more true if, to overcome the former problem, the experimental temperature range is increased towards higher T.

There are several additional reasons why the author would like to recommend a viscosity basis for ϵ/k and r_0. There is the observation for example that for *small* molecules, consisting of up to 4–6 atoms, the differences between the η-based and $B(T)$-based parameters are small and apparently random. Secondly, there are far more reliable viscosity data than $B(T)$ data available in the literature. In addition to the tabulation already mentioned [135], there are other sources to gas viscosity data such as the Landolt-Börnstein tables [136]. The exact extraction of ϵ/k and r_0 data from these is an intricate matter, but good approximations can be obtained using $r_0 = 1.13\sigma_\infty$ and $\epsilon/k = 0.906\ C$, where σ_∞ and C are constants of the Sutherland viscosity equation which are also listed in the above mentioned tables. Lastly we would like to call attention to the method developed by Stiel and Thodos [137], relating ϵ/k and r_0 to critical constants. For non-polar compounds they found empirically:

$$\epsilon/k = 65.3 \, T_c Z_c^{\,18/5} \qquad (170)$$

$$r_0 = 0.1866 \, V_c^{\,1/3} Z_c^{\,-6/5} \qquad (171)$$

where Z_c is the critical compressibility $p_c V_c/RT_c$, T_c being the critical temperature in °K, V_c the critical volume in ml/mole, p_c being in atm and r_0 being in Å. After calibrating their relation on viscosity-based data on small molecules, the authors claimed that the uncertainties involved in the use of Eqs. (170) and (171) are no worse than the variation between sources in the experimental data. This seems to be a major step forward, since many critical data are available [138, 139], including many on larger molecules, such as for example substituted benzenes and the larger alkanes, for which ϵ/k and r_0 data from η or $B(T)$ are either largely lacking or un-reliable. Closer inspection appears to reveal that again one runs into problems be-cause of large variations in experimentally determined critical constants, notably of V_c and by implication also in Z_c. In general the variations are such that the subse-quent variation in σ_w is less than 10%, but there are notable exceptions. Louman [57] quotes the example of *para*-xylene for which three sets of critical data are known. Using Eqs. (170) and (171) and then calculating σ_w of the methyl groups of *para*-xylene in CCl_4 he finds 0.381, 0.468 and 0.529 ppm, respectively. It may be true that the third source could be considered as less trustworthy, but even the 20% difference between the two former σ_w's make it desirable to find a more system-atic approach. There exist a good many empirical methods to estimate critical data from other physical data (reviewed also by Kudchacker *et al.* [139]) but one of these is much superior to all the others. It is the method for which Riedel [140] provided the back bone, but which was worked out in detail by Curl and Pitzer [141]. The method is also described in Lewis and Randall's textbook on thermodynamics [142], but it will be briefly recounted below. Three pieces of information are required; the liquid density $\rho_T(\text{g cm}^{-3})$ at some temperature T near or below the boiling point, the normal boiling point T_{760} (°K) and the temperature T_{100} (°K) at which the vapour pressure is 100 mm. T_{100} may be calculated from a vapour pressure equation for the substance such as the one used by Everett [143];

$$-{}_{10}\log p \, (\text{atm}) = 2.76 \, [(T_{760}/T_p)^{1.575} - 1] \qquad (172)$$

The calculation involves an iteration, which starts with the approximation $T_c = T_{760}/0.65$. With this approximate T_c one calculates an approximate value for the reduced temperature $T_r = T/T_c = 0.65 \, T/T_{760}$. With this T_r, a value for ρ_T/ρ_0 can be found via Table 45 and from there the quantity ρ_0 and hence the characteristic molal volume v_0 given by $v_0 = M/\rho_0 = M(\rho_T/\rho_0)/\rho_T$, M being the molecular weight.

110

Table 45. Values of ρ_T/ρ_0 as function of the reduced temperature $T_r = T/T_c$. From Curl and Pitzer [141]. Reprinted with permission of Ind. Chem. Eng.

T_r	0.3	0.4	0.5	0.6	0.7
0	0.875	0.830	0.782	0.731	0.674
1	0.870	0.826	0.777	0.726	0.668
2	0.866	0.821	0.772	0.720	0.662
3	0.862	0.816	0.767	0.715	0.656
4	0.858	0.812	0.762	0.709	0.649
5	0.853	0.807	0.757	0.703	0.642
6	0.849	0.802	0.752	0.698	0.636
7	0.844	0.797	0.747	0.692	0.629
8	0.840	0.792	0.742	0.686	0.622
9	0.835	0.787	0.737	0.680	0.615

Next the values of $\log(T_{760}/V_0)$ and of T_{100}/T_{760} are calculated and their intersept is determined from Figure 11. This yields a value for T_{760}/T_c from which an improved value for T_c is determined, plus a value for ω, which Pitzer has termed the acentric factor, because it is related to the non-central forces of a real molecule as opposed to the strict spherical symmetry and point-centralised forces assumed in the Lennard-Jones potential. The new T_{760}/T_c value is used to calculate a new T_r value and the whole cycle is repeated (usually two or three times) until the parameters T_{760}/T_c and ω do not change any more. The critical pressure p_c can then be found from Table 46.

Table 46. Dependence of p_c (atm) on T_{760}/T_c and ω calculated from $\log p_c = (\log p_c)^{(0)} + \omega(\log p_c)^{(1)}$. Reprinted with permission of Ind. Chem. Eng.

T_{760}/T_c	$(\log p_c)^{(0)}$	$(\log p_c)^{(1)}$
0.76	0.742	0.705
0.74	0.823	0.800
0.72	0.909	0.895
0.70	1.000	1.00
0.68	1.096	1.12
0.66	1.198	1.25
0.64	1.308	1.39
0.62	1.426	1.54
0.60	1.552	1.70
0.58	1.688	1.88
0.56	1.834	2.08

Finally Z_c can be calculated from $Z_c = 0.291 - 0.080\,\omega$ while V_c can be calculated using $V_c = Z_c R T_c/P_c$, where R has the value 0.082 if p_c is in atm, T_c in °K and V_c

in ml. The accuracy of this method is claimed to be as high as that of the best experimental data. Riedel, in a comparision on 26 substances found a standard deviation of 0.8% for T_c and 3.3% for p_c. No extensive comparable data for V_c are available, because of the paucity of accurate V_c data, but from the few comparisons possible the accuracy of the calculated V_c data appears to be better than $\pm$ 5% [57]. This accuracy, together with the easy accessibility of the three starting parameters ρ_T, T_{760} and T_{100} (see ref [133] or "Beilstein" or any other data collection) cou-

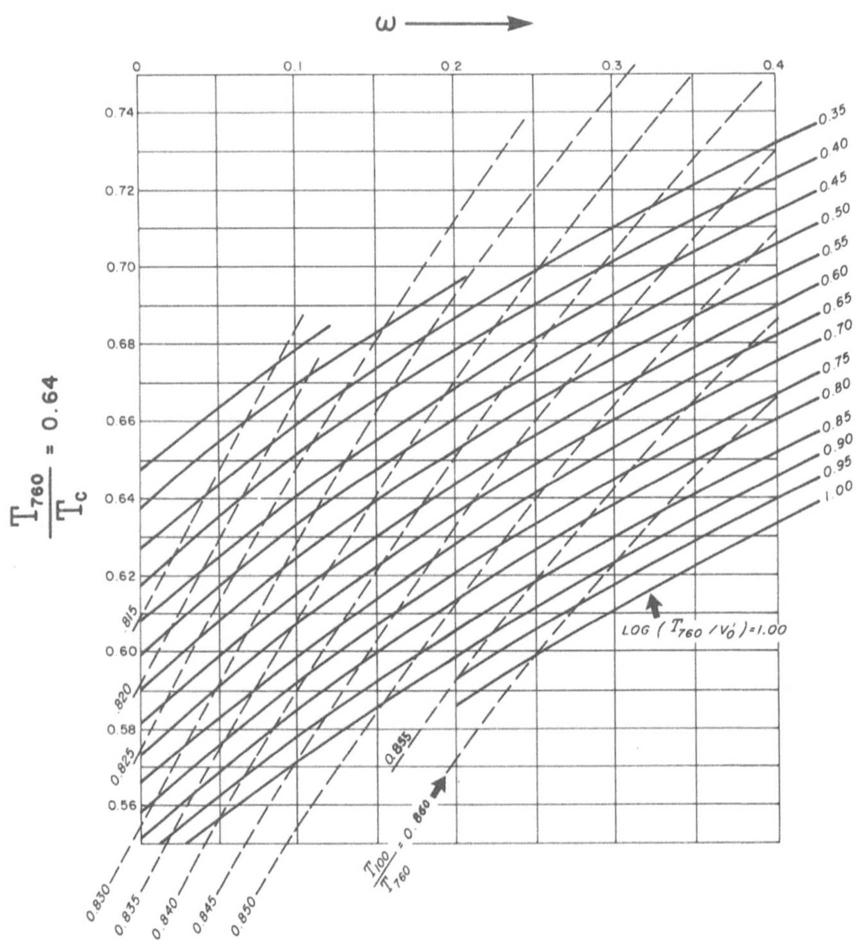

Fig. 11. T_{760}/T_c and ω as functions of T_{100}/T_{760} (steep, dotted lines) and $\log(T_{760}/V_0)$ (less steep, full lines). T_{760} is in °K and $V_0 = M/\rho_0$ is in cm^{3}mole^{-1}. By finding the point corresponding to the estimated values for T_{100}/T_{760} and $\log(T_{760}/V_0)$, the corresponding values for T_{760}/T_c and ω may be found as the ordinate and abscissa respectively of that point. From Curl and Pitzer [141]. Reproduced with permission of Ind. Eng. Chem.

pled with Eqs. (170) and (171) now provides a systematic, unambiguous and probably accurate way of estimating ϵ/k and r_0 [57, 58]. There should now be no longer any hesitancy to use the binary gas model, at least not on the pretext of unavailability of reliable potential parameters.

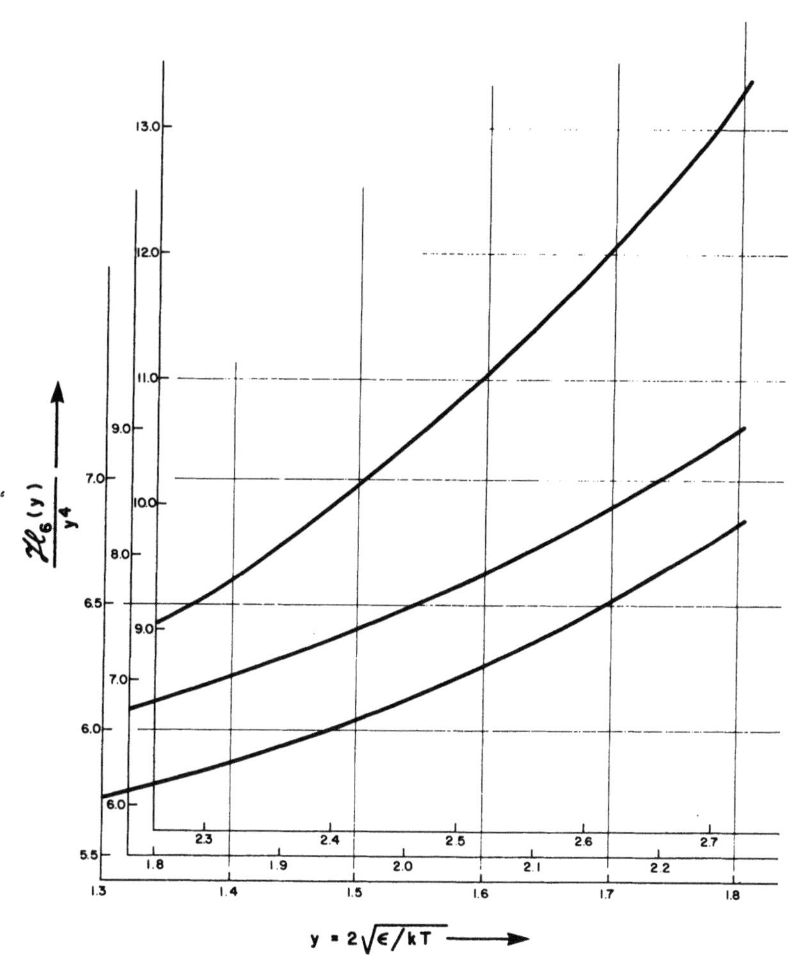

Fig. 12. $\mathscr{H}_6(y)/y^4$ as function of $y = \sqrt{\epsilon/kT}$. Calculated from the tables of Buckingham and Pople [41]

To further facilitate σ_w calculations we give in Figure 12 the function $\mathscr{H}_6(y)/y^4$ given as function of y, calculated from the tables of Buckingham and Pople [41], while in Figure 13 the site-factor S_6^g is given as function of q_0, according to Eq. (77).

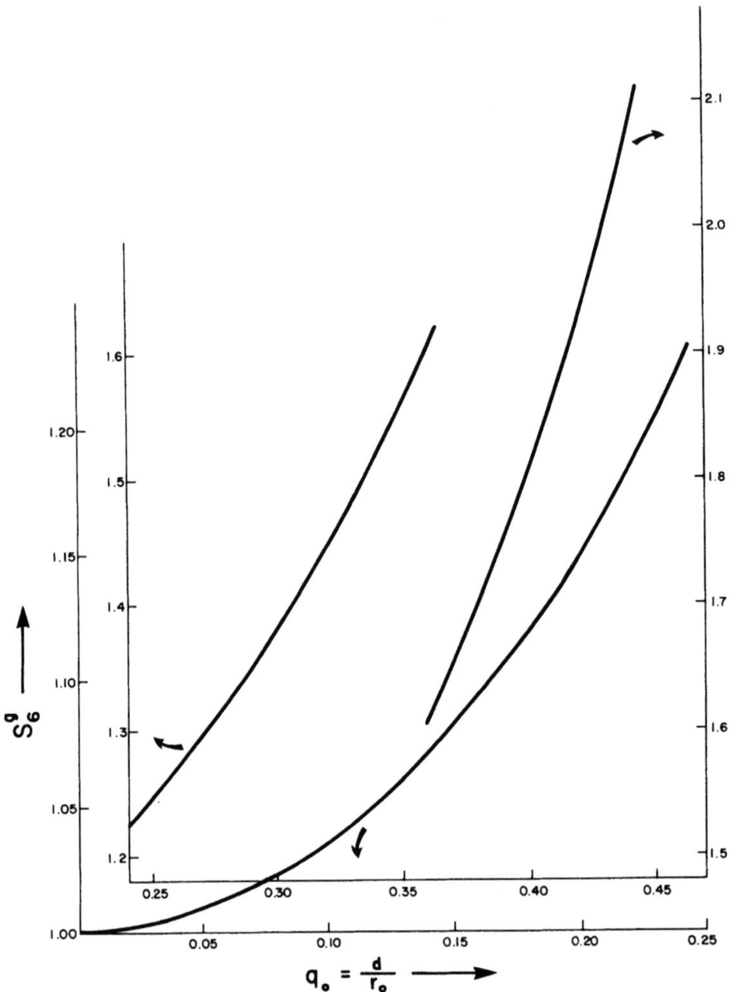

Fig. 13. Site factor S_6^g as function of $q_0 = d/r_0$, calculated from Eq. (77). Accuracy of this formula above $q_0 = 0.4$ may be doubtful (see section 6.1)

References

1. Dickinson, W. C.: Phys. Rev. **81**, 717 (1951).
2. Bothner-By, A. A., Glick, R. E.: J. Amer. Chem. Soc. **78**, 1071 (1956).
3. Bothner-By, A. A., Glick, R. E.: J. Chem. Phys. **26**, 1647 (1957).
4. Bothner-By, A. A., Glick, R. E.: J. Chem. Phys. **26**, 1651 (1957).
5. Glick, R. E., Ehrenson, S. J.: J. Chem. Phys. **29**, 459 (1958).
6. Glick, R. E., Kates, D. F., Ehrenson, S. J.: J. Chem. Phys. **31**, 567 (1959).
7. Bothner-By, A. A.: J. Mol. Spectroscopy **5**, 52 (1960).
8. Linder, B.: J. Chem. Phys. **33**, 668 (1960).
9. Marshall, T. W., Pople, J. A.: Mol. Phys. **1**, 199 (1958).
10. Stephen, M. J.: Mol. Phys. **1**, 223 (1958).
11. Buckingham, A. D., Stephen, M. J.: Trans. Far. Soc. **53**, 884 (1957).
12. Schneider, W. G., Bernstein, H. J., Pople, J. A.: J. Chem. Phys. **28**, 601 (1958).
13. American Petroleum Institute Research Project 44, "Selected Values of Properties of Hydrocarbons", N.B.S. Washington (D.C.).
14. Evans, D. F.: Proc. Chem. Soc. **1958**, 115.
15. Raynes, W. T., Buckingham, A. D., Bernstein, H. J.: J. Chem. Phys. **36**, 3481 (1962).
16. Rummens, F. H. A., Bernstein, H. J.: J. Chem. Phys. **43**, 2971 (1965).
17. Rummens, F. H. A., Raynes, W. T., Bernstein, H. J.: J. Chem. Phys. **72**, 2111 (1968).
18. Buckingham, A. D., Schaefer, T., Schneider, W. G.: J. Chem. Phys. **32**, 1227 (1960).
19. Barter, C., Meisenheimer, R. G., Stevenson, D. P.: J. Phys. Chem. **64**, 1312 (1960).
20. Broersma, S.: J. Chem. Phys. **17**, 873 (1949).
21. "Tables de Constantes et Données Numériques", Pergamon Press, Vol. 7 (1957).
22. Landolt-Börnstein: "Zahlenwerten und Funktionen", 6 ed., Band II, Teil 10, Springer Verlag (1967).
23. Gordon, S., Dailey, B. P.: J. Chem. Phys. **34**, 1084 (1961).
24. Rummens, F. H. A.: Mol. Phys. **21**, 535 (1971).
25. Linder, B.: J. Chem. Phys. **37**, 963 (1962).
26. Howard, B. B., Linder, B., Emerson, M.T.: J. Chem. Phys. **36**, 485 (1962).
27. Lumbroso, N., Wu, T. K., Dailey, B. P.: J. Phys. Chem. **67**, 2469 (1963).
28. Linder, B.: J. Chem. Phys. **35**, 371 (1961).
29. Barriol, J., De Montgolfier, Ph.: Compt. Rend. Acad. Sci. Paris **262** C, 1638 (1966).
30. De Montgolfier, Ph.: J. Chim. Phys. **64**, 639 (1967).
31. De Montgolfier, Ph.: Compt. Rend. Acad. Sci. Paris **263** C, 505 (1966).
32. Barriol, J., Weissbecker, A.: Compt. Rend. Acad. Sci. Paris **259** C, 2831 (1964).
33. De Montgolfier, Ph.: J. Chim. Phys. **66**, 685 (1969).
34. Rummens, F. H. A.: J. Amer. Chem. Soc. **92**, 3214 (1970);
35. Chenon, M.-T.: Ph. D. Thesis, Faculté des Sciences, Université de Paris (1970).
36. Chenon, M.-T, Bouquant, J, Lumbroso-Bader, N.: J. Chim. Phys. **67**, 1252 (1970).
37. McRae, E. G.: J. Phys. Chem. **61**, 562 (1957).
38. Laszlo, P., Speert, A., Raynes, W. T.: J. Chem. Phys. **51**, 1677 (1969).
39. Bayliss, N. S.: J. Chem. Phys. **18**, 292 (1950).
40. Buckingham, A. D., Pople, J. A.: Disc. Faraday Soc. **22**, 17 (1956).
41. Buckingham, A. D., Pople, J. A.: Trans. Faraday Soc. **51**, 1173 (1955).
42. Petrakis, L., Bernstein, H. J.: J. Chem. Phys. **38**, 1562 (1963).
43. Bernstein, H. J., Raynes, W. T.: presented at the NMR Symposium, Boulder (Colorado), July 1962.
44. Kromhout, R. A., Linder, B.: J. of Magn. Resonance **1**, 450 (1969).
45. Marshall, T. W., Pople, J. A.: Mol. Phys. **3**, 399 (1960).
46. Linder, B., Hoernschemeyer, D.: J. Chem. Phys. **46**, 784 (1967).
47. Dayan, E., Widenlocher, G.: Compt. Rend. Acad. Sci. Paris **263** B, 1346 (1966).

48. Dayan, E.: Thesis, Faculté des Sciences, Université de Paris (1965).
49. Dayan, E., Widenlocher, G.: Compt. Rend. Acad. Sci. Paris **257B**, 883 (1963).
50. Oldenziel, J. G.: Ph. D. Thesis, University of Amsterdam (1972)
51. Meinzer, R. A.: Ph. D. Thesis (No. 66–4238, University Microfilm Inc., Ann Arbor), University of Illinois (1965).
52. Mohanty, S., Bernstein, H.J.:
    (a) Chem. Phys. Lett. **4**, 575 (1970).
    (b) J. Chem. Phys. **54**, 2254 (1971).
53. Jouve, P: Thesis, Faculté des Sciences, Université de Paris (1966).
54. Friedrich, H. J.: Zeitschrift für Naturforschung **19b**, 663 (1964), **20b**, 1021 (1965).
55. Raza, M. A.: Ph. D. Thesis, Univ. of Sheffield (1969).
56. Raza, M. A., Raynes, W. T.: Mol. Phys. **19**, 199 (1970).
57. Louman, F. J. A.: Ph. D. Thesis, Univ. of Sask, Regina Campus (1972).
58. Rummens, F. H. A., Louman, F. J. A.: J. Magn. Res. Resonance **8**, 332 (1972).
59. Raynes, W. T., Raza, M. A.: Mol. Phys. **17**, 157 (1969).
60. Chaigneau, M., Dayan, E., Widenlocher, G.: Compt. Rend. Acad. Sci. Paris **255**, 2597 (1962).
61. Hindermann, D. K., Cornwell, C. D.: J. Chem. Phys. **48**, 2017 (1968).
62. Rummens, F. H. A., De Meyer, H. J.:
    (a) unpublished results (1973).
    (b) Vth International Symposium on Magnetic Resonance, Bombay (India) (1974).
63. Raynes, W. T.: Mol. Phys. **17**, 169 (1969).
64. Pitzer, K. S.: J. Amer. Chem. Soc. **77**, 3427 (1955).
65. Hirschfelder, J. O., Linnett, J. W.: J. Chem. Phys. **18**, 130 (1950).
66. Jameson, A. K., Jameson, C. J., Gutowsky, H. S.: J. Chem. Phys. **53**, 2310 (1970).
67. Adrian, F. J.: Phys. Rev. **136**, A980 (1964).
68. Buckingham, A. D., Lawley, K. P.: Mol. Phys. **3**, 219 (1960).
69. Michels, A., De Boer, J., Bijl, A.: Physica **4**, 981 (1937).
70. Yamada, H.: Chemistry Letters **9**, 747 (1972).
71. Yamada, H., Ishihara, T., Kinugasa, T.: J. Amer. Chem. Soc. **96**, 1935 (1974).
72. Yamada, H.: Rev. Sci. Instrum. **45** (1974).
73. Von Jouanne, J.: Ph. D. Thesis, Univ. Frankfurt am Main (1971).
74. Von Jouanne, J., Heidelberg, J.: J. Magn. Res. **7**, 1 (1972).
75. Margenau, H.: J. Chem. Phys. **6**, 897 (1938).
76. Yonemoto, T: Can. J. Chem. **44**, 223 (1966).
77. Buckingham, A. D.: Can. J. Chem. **38**, 300 (1960).
78. Streever, R. L., Carr, H. Y.: Phys. Rev. **121**, 20 (1961).
79. Hunt, E. R., Carr, H. Y.: Phys. Rev. **130**, 2302 (1963).
80. Kanegsberg, E., Pass, B., Carr, H. Y.: Phys. Rev. Letters **23**, 572 (1969).
81. De Boer, J., Van Leeuwen, J. M. J., Groeneveld, J.: Physica **30**, 2265 (1964).
82. Petrakis, L., Sederholm, C. H.: J. Chem. Phys. **35**, 1174 (1961).
83. Buckingham, A. D.: J. Chem. Phys. **36**, 3096 (1962).
84. Raynes, W. T., Davies, A. M., Cook, D. B.: Mol. Phys. **21**, 123 (1971).
85. Jameson, A. K., Jameson, C. J.: J. Amer. Chem. Soc. **95**, 8559 (1973).
86. Mourits, F. M., Rummens, F. H. A.: Regina, unpublished work (1974).
87. Mohanty, S., Bernstein, H. J.: J. Magn. Resonance **8**, 152 (1972).
88. Spanier, R. F., Malinowski, E. R.: J. Chem. Phys. **47**, 1560 (1967)
89. Spanier, R. R., Vladimiroff, T., Malinowski, E. R.: J. Chem. Phys. **45**, 4355 (1966).
90. Malinowski, E. R., Weiner, P. H.: J. Amer. Chem. Soc. **92**, 4193 (1970).
91. Weiner, P. H., Malinowski, E. R., Levinstone, A. R.: J. Phys. Chem. **74**, 4537 (1970).
92. Weiner, P. H., Malinowski, E. R.: J. Phys. Chem. **75**, 1207 (1971).
93. Bernstein, H. J.: Pure and Applied Chemistry **32**, 79 (1972).
94. Rummens, F. H. A.: J. Mol. Phys. **19**, 423 (1970).
95. Homer, J.: Tetrahedron **23**, 4065 (1967).
96. Schug, J. C.: J. Chem. Phys. **70**, 1816 (1966).
97. Abraham, R. J.: Mol. Phys. **4**, 369 (1961).
98. Abraham, R. J., Wileman, D. F., Bedford, C. R.: J. Chem. Soc., Perkin II, **1973**, 1027.

99. Abraham, R. J., Wileman, D. F.: J. Chem. Soc. Perkin II, **1973**, 1522.
100. Bacon, M. R., Maciel, G. E.: J. Amer. Chem. Soc. **95**, 2413 (1973).
101. Glick, R. E., Ehrenson, S. J.: J. Phys. Chem. **62**, 1599 (1958).
102. Evans, D. F.: J. Chem. Soc. **1960**, 877.
103. Raynes, W. T., Raza, M. A.: Mol. Phys. **20**, 555 (1971).
104. Frei, K., Bernstein, H. J.: unpublished results (1961), (private communication).
105. Cyr, N., Raza, M. A., Reeves, L. W.: Mol. Phys. **24**, 459 (1972).
106. Emsley, J. W., Phillips, L.: Mol. Phys. **11**, 437 (1966).
107. ASTM standard E 368 proposal of ASTM subcommittee E 13.07, American Standard for Testing and Materials, 1916 Race St., Philadelphia (Pa), 19103.
108. Bacon, M. R., Maciel, G. E., Musker, W. K., Scholl, R.: J. Amer. Chem. Soc. **93**, 2537(1971).
109. Scholl, R. L., Maciel, G. E., Musker, W. K.: J. Amer. Chem. Soc. **94**, 6376 (1972).
110. Heckmann, G., Fluck, E.: Mol. Phys. **23**, 175 (1972).
111. Heckmann, G., Fluck, E.: Naturforschung **B 24**, 1092 (1969).
112. Heckmann, G., Fluck, E.: Naturforschung **B 26**, 63 (1971).
113. Brinkmann, D.: Helv. Phys. Acta **36**, 431 (1963).
114. Brinkmann, D.: Phys. Rev. Letters **13**, 187 (1964).
115. Jameson, C. J., Jameson, A. K.: Mol. Phys. **28**, 957 (1971).
116. Buckingham, A. D., Kollman, P. A.: Mol. Phys. **23**, 65 (1972).
117. Becconsall, J. K., Daves, G. D., Anderson, W. R.: J. Amer. Chem. Soc. **92**, 430 (1970).
118. Rummens, F. H. A., Krystynak, R. H.: J. Amer. Chem. Soc. **94**, 6914 (1972).
119. The Wilmad Glass Company, Oak Road, (N.J.) 08310, U.S.A.
120. Zimmerman, J. R., Foster, M. R.: J. Phys. Chem. **61**, 282 (1957).
121. Smith, G. W., "Compilation of Diamagnetic Susceptibilities of Organic Compounds", General Motors Research publication GMR-317, G. M. Corp., Detroit (Mich), (1959).
122. Lussan, C.: J. Chim. Phys. **61**, 462 (1964).
123. Frei, K., Bernstein, H. J.: J. Chem. Phys. **37**, 1891 (1962).
124. Vincent, E. J., Phan-Tan-Lun, R., Metzer, J., Surzur, J. M.: Compt. Rend. Acad. Sci. Paris **260**, 6345 (1965).
125. Morin, M. G., Paulett, G., Hobbs, M. E.: J. Phys. Chem. **60**, 1594 (1956).
126. Douglass, D. C., Fratiello, A.: J. Chem. Phys, **39**, 3161 (1963).
127. Spanier, R. F., Vladimiroff, T., Malinowski, E. R.: J. Chem. Phys. **45**, 4355 (1960).
128. Williams, G. A., Gutowsky, H. S.: J. Chem. Phys. **25**, 128 (1956).
129. Malinowski, E. R., Pierpaoli, A. R.: J. Magn. Resonance **1**, 509 (1969).
130. Homer, J., Whitney, P. M.: J. Chem. Soc. Chem. Comm. **1972**, 153.
131. Frost, D. J., Hall, G. E.: Mol. Phys. **10**, 191 (1965).
132. Van Geet, A. L.: Anal. Chem. **40**, 2227 (1968), *ibid 42*, 679 (1970).
133. (a) Landolt-Börnstein, „Zahlenwerte und Funktionen", Band III, Teil **2**, Springer Verlag (1960).
    (b) "Advances in Chemistry" ASTM Monograph Series Nos. 15 (1955), 22 (1959), 29 (1961), American Chemical Society, Washington (D.C.).
    (c) "Handbook of Chemistry and Physics", The Chemical Rubber Publishing Co..
    (d) "International Critical Tables" Volume III.
    (e) Timmermans, J., "Physico-Chemical Constants of Pure Organic Compounds", Elsevier Publishing Co., Vol. 1 (1950), Vol. 2 (1965).
134. Kiser, R. W., "Table of Ionization Potentials", U.S. Atomic Energy Commission, Kansas State University, (1960).
135. Hirschfelder, J. O., Curtiss, C. F., Bird, R. B.: "Molecular Theory of Gases and Liquids", John Wiley & Sons, Inc. (New York), 2nd Edition (1964) 1110, 1212.
136. Landolt-Börnstein, „Zahlenwerte und Funktionen", 6th Ed., Vol. 1, Part 1.
137. Stiel, L. T., Thodos, G.: J. Chem. Eng. Data 7, 24 (1962), J. Amer. Inst. Chem. Eng. **10**, 266 (1964).
138. "American Institute of Physics Handbook", 2nd Ed., American Institute of Physics Inc., New York (1963).
139. Kudchacker, A. P., Alani, G. H., Zwolinski, B. J.: Chem. Rev. 68 , 659 (1968).

References

140. Riedel, L.: Chem. Ing. Techn. **27**, 475 (1955).
141. Curl, R. F., Pitzer, K. S.: Ind . Eng. Chem. **50**, 265 (1958).
142. Lewis, G. N., Randall, M.: "Thermodynamics", revised by K. S. Pitzer and L. Brewer, McGraw-Hill Book Co. Inc., New York (1961).
143. Everett, D. H.: J. Chem. Soc. **1960**, 2566.

Author Index Volumes 1–9

NMR

Basic Principles and Progress
Grundlagen und Fortschritte

Editors: P. Diehl, E. Fluck, R. Kosfeld

Volume 7

C. W. HILBERS, C. MACLEAN:

**NMR of Molecules Oriented
in Electric Fields**

The Spin Hamiltonian of Molecules Partially Aligned by an External Electric Field. Examples. Experiments. Some Model Calculations of the Alignment. Statistical Calculations of the Alignment. Comparison between the Kerr Effect and the NMR Electric Field Effect. Concluding Remarks.

H. PFEIFFER:

**Nuclear Magnetic Resonance
and Relaxation of Molecules
Adsorbed on Solids**

Nuclear Magnetic Resonance and Relaxation — Basic Principles and Qualitative Discussion. Results of Nuclear Magnetic Relaxation Theory — Relaxation Analysis, Diffusion Analysis. Nuclear Magnetic Relaxation and Diffusion Study of Some Characteristic Systems.

56 figures. V, 153 pages. 1972
ISBN 3-540-05687-4

Volume 8

C. RICHARD, P. GRANGER:

**Chemically Induced Dynamic Nuclear and Electron Polarizations —
CIDNP and CIDEP**

Origin of the CIDNP Effects. The Theory of the CIDNP Effect. Applications to the Study of Chemical Reactions and Magnetic Properties. The Chemically Induced Dynamic Electron Polarization (CIDEP Effect).

26 figures. II, 127 pages. 1974
ISBN 3-540-06618-7

Volume 9

C. L. KHETRAPAL,
A. C. KUNWAR, A. S. TRACEY,
P. DIEHL:

Lyotropic Liquid Crystals

Nuclear Magnetic Resonance Studies in Lyotropic Liquid Crystals: Introduction. Studies of Lyotropic Liquid Crystals. Studies of Molecular and Ionic Species Dissolved in the Nematic Phase of Lyotropic Liquid Crystals.

18 figures. 3 tables. IV, 85 pages. 1975
ISBN 3-540-07303-5

Springer-Verlag
Berlin
Heidelberg
New York

Structure and Bonding

Editors: J. D. Dunitz, P. Hemmerich,
R. H. Holm, J. A. Ibers, C. K. Jørgensen,
J. B. Neilands, D. Reinen, R. J. P. Williams

Volume 19
Chemical Bonding in Solids
37 figures. IV, 165 pages. 1974
ISBN 3-540-06908-9

Contents: R. D. Shannon, H. Vincent:
Relationship between Covalency, Interatomic Distances, and Magnetic Properties in Halides and Chalcogenides. –
A. Kjekshus, T. Rakke: Considerations on
the Valence Concept. – A. Kjekshus,
T. Rakke: Geometrical Considerations on
the Marcasite Type Structure. –
G. C. Allen, K. D. Warren: The Electronic
Spectra of the Hexafluoro Complexes of
the Second and Third Transition Series.

Volume 20
Biochemistry
57 figures. IV, 167 pages. 1974
ISBN 3-540-07053-2

Contents: A. S. Mildvan, C. M. Grisham:
The Role of Divalent Cations in the
Mechanism of Enzyme Catalyzed Phos-
phoryl and Nucleotidyl Transfer Reac-
tions. – H. P. C Hogenkamp, G. N. Sando:
The Enzymatic Reduction of Ribo-
nucleotides. – W. T. Oosterhuis: The
Electronic State of Iron in Some Natural
Iron Compounds: Determination by
Mössbauer and ESR Spectroscopy. –
A. Trautwein: Mössbauer-Spectroscopy
on Heme Proteins.

Volume 21
**Recent Impact of Physics on Inorganic
Chemistry**
62 figures. IV, 144 pages. 1975
ISBN 3-540-07109-1

Contents: B. C. Tofield: The Study of
Covalency by Magnetic Neutron Scatter-
ing. – B. Fricke: Superheavy Elements.

Volume 22
Rare Earths
36 figures. IV, 172 pages. 1975
ISBN 3-540-07268-3

Contents: E. Nieboer: The Lanthanide
Ions as Structural Probes in Biological
and Model Systems. – C. K. Jørgensen:
Partly Filled Shells Constituting Anti-
bonding Orbitals with Higher Ionization
Energy than their Bonding Counterparts.
– R. D. Peacock: The Intensities of
Lanthanide f ↔ f Transitions. –
R. Reisfeld: Radiative and Non-Radiative
Transitions of Rare-Earth Ions in
Glasses.

Springer-Verlag
Berlin
Heidelberg
New York